Fluorescence Spectroscopy

New Methods and Applications

Edited by Otto S. Wolfbeis

With 176 Figures

Springer-Verlag

Berlin Heidelberg New York
London Paris Tokyo
Hong Kong Barcelona Budapest

Professor Dr. OTTO S. WOLFBEIS

Karl Franzens University
Institute of Organic Chemistry
Analytical Division
Heinrichstraße 28
A-8010 Graz
Austria

ISBN 3-540-55281-2 Springer-Verlag Berlin Heidelberg New York
ISBN 0-387-55281-2 Springer-Verlag New York Berlin Heidelberg

Library of Congress Cataloging-in-Publication Data
Fluorescence spectroscopy : new methods and applications / Edited by Otto S. Wolfbeis.
p. cm. Includes bibliographical references and index. ISBN 3-540-55281-2 (alk. paper). --
ISBN 0-387-55281-2 (alk. paper) 1. Fluorescence spectroscopy--Congresses. I. Wolfbeis,
Otto S. [DNLM: 1. Spectrometry, Fluorescence--methods--congresses. QH 324.9 F4 F646]
QH 324.9. F4 F56 1992 574.19'285--dc20 DNLM/DLC for Library of Congress 92-2436 CIP

© Springer-Verlag Berlin Heidelberg 1993
Printed in Germany

Typesetting: Macmillan, Bangalore, India;
Offsetprinting: Mercedes Druck, Berlin; Binding: Lüderitz & Bauer, Berlin
52/3020- 5 4 3 2 1 0 – Printed on acid-free paper

Although the phenomenon of luminescence has been known since the earliest times, fluorescence (in its present meaning) has been described relatively recently, because a differentiation between phosphorescence, fluorescence, and chemi- and bio-luminescence as well as other forms of luminescence was not made until the 19th century [1]. All of these phenomena have meanwhile resulted in the development of extremely important methods which are widely applied in many scientific fields, but particularly in analytical and clinical chemistry, in biochemistry, biophysics and biology.

The basic work on fluorescence spectroscopy and its phenomena was performed by physicists. A standard book on the fluorescence of organic compounds was written by Förster [2] in 1951 but unfortunately was never translated into English. It is still an invaluable source of information. Other books on basic aspects of fluorescence spectroscopy include those of Parker [3] and Becker [4]. It required quite some time to transfer fluorescence technology to the fields where it has experienced its greatest success, viz. in analytical and biosciences. The transfer from merely academic research to practicability has been greatly accelerated by the success of early books including the classical treatises of Udenfriend [5], Guilbault [6], and others. Within a period of some 30 years, fluorescence spectroscopy has come to occupy a prominent position in various fields. This is reflected by a number of books on the topic including those of Chen and Edelhoch [7], Wehry [8], Zander [9], Lakowicz [10], Steiner [11], Schulman [12], Goldberg [13], and the more recent ones by Dewey [14] and Baeyens et al. [15]. In addition, there exists a body of reviews and chapters contained in other books where exciting progress in selected areas of applied fluorescence spectroscopy is described.

In looking at the literature cited, we can see fluorescence spectroscopy gradually moving from physics to chemistry and biology. One also notes that books on fluorescence spectroscopy frequently are no longer written by a single author, but rather are edited by authors with some expertise in selected areas, while many (or all) chapters are being written by experts in the respective fields. This certainly reflects the fact that fluorescence spectroscopy is a truly interdisciplinary science.

In view of the increasing significance of fluorescence spectroscopy to biochemical, biophysical and analytical sciences, and given the fact that this kind of spectroscopy is experiencing tremendous progress at present, it appeared beneficial to review the state of the art and to demonstrate the extreme utility of fluorometry in very different areas. At present, the significance of fluorescence spectroscopy to chemical, biochemical and biophysical sciences is probably second only to NMR spectroscopy. Like the latter – another brainchild of physics – it has found its greatest utility in fields other than physics.

Applied fluorescence spectroscopy is highly interdisciplinary: Organic chemists are the ones who design and characterize synthetic fluorophors which – in turn – are used in other areas including immunoassay, membrane studies, cell sorting and tracer studies, to mention only a few. On the other hand, new methods are being developed that make use of sophisticated new components including (diode) lasers and LEDs, fiber optics, video cameras, fast optical imaging devices and techniques, data loggers, and – of course – intelligent software. If one has to mention one or two single factors for the tremendous increase in the popularity of fluorescence spectroscopy in the 1980s, it probably is the commercial availability of tailor-made and specific fluorescent probes, along with the success of fluorescence lifetime measurements. The commercially most successful single area of applied fluorometry definitely is luminescence immunoassay which is in the process of almost completely replacing radio-immunoassay. This again has resulted in a wide acceptance and awareness of fluorescence techniques.

This book comprises the plenary lectures of a 3-day conference on "Methods and Applications of Fluorescence Spectroscopy", held in Graz, Austria, in October 1991. It covers various areas where fluorescence spectroscopy is applied, but since the greatest progress has definitely been made in analytical, biomedical and bio-physical sciences this is reflected by the contents of the book. The book is divided into sections on 1. New Methods, 2. New Applications, 3. Fluorimetric Analysis, 4. Fluorescence Immunoassay, and 5. Fluorescence in Biomedical Sciences. In each of these areas, impressive progress has been made in the past few years. In an introductory chapter, a report is given on the status and on current trends in analytical fluorescence.

Part 1 (New Methods) starts with an exciting chapter on fluorescence correlation spectroscopy. Nothing documents the progress in methodology better than the report on the feasibility of detecting *one single molecule* by fluorescence correlation spectroscopy as described in Chapter 2. C. D. MacKay in Chapter 3 reports on the progress that has been made in the very important field of fast optical imaging techniques, while in Chapter 4, kinetic studies on fluorescent probes using synchrotron radiation are treated in some detail. Then, A.J.W.G. Visser describes how relaxation spectroscopy can be used to study the dynamics of proteins. D. Oelkrug reports on fluorescence spectroscopic studies of light scattering materials (which are more frequently encountered in laboratory practice than clear and homogeneous solutions).

The section on New Applications starts with a chapter on oxygen sensor which exploit the phenomenon of dynamic fluorescence quenching. Because oxygen sensors can be coupled to all kinds of biochemical reactions during which oxygen is consumed or produced, such sensors are potentially useful in optical biosensors as well. In the following chapters, Schneckenburger and Schmidt describe how fluorescence spectrometry can be applied for early detection of forest decline, while H. Riegler et al. report on fluorescence microscopy of surfactant monolayers.

Moving on to fluorescence lifetime studies, J. R. Lakowicz gives an excellent overview of the recent progress that has been made with respect to biomedical applications of fluorescence lifetime imaging, and A. Hermetter reports on the incorporation of ether phospholipids into membranes and how this can be applied for membrane studies using both phase and steady-state fluorometry. P. K. J. Kinnunen uses pyrene-labelled lipids as fluorescent probes for lifetime studies on biomembranes and membrane models. Both papers reflect the tremendous success that fluorescence spectroscopy has had in the area of membrane biochemistry and biophysics. Slavik uses old and new types of fluorescent probes for optical detection of intracellular ion concentrations using fluorescence microscopy and related techniques to gain insight into biological events on a cellular level that would not have been considered possible some 20 years ago.

Analytical applications of fluorimetry are treated in Part 3. Semiconductor laser fluorometry (with its obvious advantages over other kinds of laser fluorometry, not to mention conventional fluorometry) is reviewed by J.N. Miller. The diode laser which is very small and can be battery-powered is, in fact, an outgrowth from CD-player technology, but has now found its way into spectroscopy, because long-wave absorbing fluorescent probes have been developed. M. Valcarcel applies luminescence detection in flow injection analysis (FIA), a technique that has enjoyed enormous (commercial) success over the last 10 years and has substantially simplified routine analysis. Aspects of environmental analysis using multi-dimensional fluorescence spectroscopy, with particular emphasis given to hydrological studies, are described by M. Goldberg.

Part 4 comprises three chapters on fluorescence immunoassay. Ch. Klein from Boehringer Mannheim reviews fluorescence polarization immunoassay, I. Hemmilä reports on progress in delayed fluorescence immuno-assay, and E. Jansen on chemiluminescence immunoassay and its applications to environmental monitoring.

Part 5 on Fluorescence in Biomedical Sciences begins with a review by W.N. Ross on optical measurement of calcium and voltage transients, a technique that has provided tremendous insight into molecular processes going on during signal transport in biological matter. New results obtained with voltage-sensitive dyes and their application in studies on the brain stem are presented by K. Kamino et al.

I am sure that this book represents a substantial and representative fraction of current activity in this exciting field, and truly hope that this will contribute further to the popularity and success of fluorescence spectroscopy.

Graz, September 1992   Otto S. Wolfbeis

References

1. Harvey E N (1957) A history of luminescence from the earliest times until 1900. Am. Phil. Soc., Philadelphia
2. Förster Th (1951) Fluorescenz organischer Verbindungen. Vandenhoeck and Rupprecht, Göttingen
3. Parker C A (1968) Photoluminescence of solutions. Elsevier, Amsterdam
4. Becker R S (1969) Theory and interpretation of fluorescence and phosphorescence. Wiley Interscience, New York
5. Udenfriend S (1962/1971) Fluorescence assay in biology and medicine. Academic, New York, vol 1 (1962) and 2 (1967)
6. Guilbault C G (1973) Practical fluorescence: Theory, methods, and techniques. Marcel Dekker, New York
7. Chen R F, Edelhoch H (1976) Biochemical fluorescence: Concepts. Marcel Dekker, New York, vols 1 and 2
8. Wehry E L (ed) (1981) Modern fluorescence spectroscopy. Plenum, New York
9. Zander M (1981) Fluorimetrie. Springer, Berlin Heidelberg New York
10. Steiner R F (ed) (1983) Excited states of biopolymers. Plenum, New York
11. Lakowicz J R (1983) Principles of fluorescence spectroscopy. Plenum, New York
12. Schulman S G (ed) (1985/1988) Molecular luminescence spectroscopy. Methods and applications. John Wiley, New York, vols 1 (1985) and 2 (1988)
13. Goldberg M C (ed) (1989) Luminescence applications in biological, chemical environmental and hydrological sciences. ACS Symposium Ser. vol 383, Am Chem Soc, Washington, DC
14. Dewey T G (ed) (1991) Biophysical and biochemical aspects of fluorescence spectroscopy. Plenum, New York
15. Baeyens W R G, De Keukelaire D, Korkidis K (eds) (1991) Luminescence techniques in chemical and biochemical analysis. Marcel Dekker, New York

Contents

Part 2. New Applications of Fluorometry

Part 4. Fluorescence Immunoassay

List of Contributors

Ph. I. H. Bastiaens, Department of Biochemistry, Agricultural University, Dreijenlaan 3, NL-6703 HA Wageningen, The Netherlands

H.-G. Batz, Boehringer Mannheim GmbH, Werk Tutzing, Bahnhofstrasse 9-15, D-8132 Tutzing, Germany

K. Berndt, University of Maryland, School of Medicine, Center for Fluorescence Spectroscopy, Department of Biological Chemistry, 660 West Redwood Street, Baltimore, MD 21201, USA

M. B. Brown, Department of Chemistry, Loughborough University of Technology, Loughborough, Leicestershire LE11 3TU, UK

M. Brun, Institut für Physikalische und Theoretische Chemie der Universität Tübingen, Auf der Morgenstelle 8, D-7400 Tübingen, Germany

B. Draeger, Boehringer Mannheim GmbH, Werk Tutzing, Bahnhofstrasse 9-15, D-8132 Tutzing, Germany

M. C. Goldberg, U.S. Geological Survey, P.O. Box 25046 MS 424, Denver Federal Center, Lakewood, CO 80225, USA

H.-J. Guder, Boehringer Mannheim GmbH, Werk Tutzing, Bahnhofstrasse 9-15, D-8132 Tutzing, Germany

R. Günther, Institut für Physikalische und Theoretische Chemie der Universität Tübingen, Auf der Morgenstelle 8, D-7400 Tübingen, Germany

I. Hemmilä, Wallac Biomedical Laboratory, P.O. Box 10, SF-20101 Turku 10, Finland

A. Hermetter, Institut für Biochemie und Lebensmittelchemie, Technische Universität Graz, Petersgasse 12/II, A-8010 Graz, Austria

R. Herrmann, Boehringer Mannheim GmbH, Werk Tutzing, Bahnhofstrasse 9-15, D-8132 Tutzing, Germany

A. Hirota, Department of Physiology, Tokyo Medical and Dental University, School of Medicine, 5-45, Yushima 1-chome, Bunkyo-ku, Tokyo 113, Japan

E. H. J. M. Jansen, Laboratory for Toxicology, National Institute of Public Health and Environmental Protection, P.O. Box 1, NL-3720 BA Bilthoven, The Netherlands

M. L. Johnson, University of Virginia, School of Medicine, Department of Pharmacology, Charlottesville, VA 22908, USA

H.-P. Josel, Boehringer Mannheim GmbH, Werk Tutzing, Bahnhofstrasse 9-15, D-8132 Tutzing, Germany

E. Kalb, Institut für Biochemie und Lebensmittelchemie, Technische Universität Graz, Petersgasse 12//II, A-8010 Graz, Austria

K. Kamino, Department of Physiology, Tokyo Medical and Dental University, School of Medicine, 5-45, Yushima 1-chome, Bunkyo-ku, Tokyo 113, Japan

P. K. J. Kinnunen, Lipid Research Laboratory, Department of Medical Chemistry, University of Helsinki, Siltavuorenpenger 10, SF-00170 Helsinki, Finland

Ch. Klein, Boehringer Mannheim GmbH, Werk Tutzing, Bahnhofstrasse 9-15, D-8132 Tutzing, Germany

H. Komuro, Department of Physiology, Tokyo Medical and Dental University, School of Medicine, 5-45, Yushima 1-chome, Bunkyo-ku, Tokyo 113, Japan

A. Koiv, Lipid Research Laboratory, Department of Medical Chemistry, University of Helsinki, Siltavuorenpenger 10, SF-00170 Helsinki, Finland

J. R. Lakowicz, University of Maryland, School of Medicine, Center for Fluorescence Spectroscopy, Department of Biological Chemistry, 660 West Redwood Street, Baltimore, MA 21201, USA

J. Loidl, Institut für Biochemie und Lebensmittelchemie, Technische Universität Graz, Petersgasse 12//II, A-8010 Graz, Austria

M. D. Luque de Castro, Department of Analytical Chemistry, Faculty of Sciences, University of Córdoba, 14004 Córdoba, Spain

C. D. MacKay, Astromed Ltd., Innovation Centre, Cambridge Science Park, Milton Road, Cambridge, CB4 4GS, UK

U. Mammel, Institut für Physikalische und Theoretische Chemie der Universität Tübingen, Auf der Morgenstelle 8, D-7400-Tübingen, Germany

Ü. Mets, Karolinska Institute, Department of Medical Biophysics, Box 60400, S-10401 Stockholm, Sweden

J. N. Miller, Department of Chemistry, Loughborough University of Technology, Loughborough, Leicestershire LE11 3TU, UK

H. Möhwald, Institut für Physikalische Chemie, Universität Mainz, Welder-Weg 11, D-6500 Mainz, Germany

Y. Momose-Sato, Department of Physiology, Tokyo Medical and Dental University, School of Medicine, 5-45, Yushima 1-chome, Bunkyo-ku, Tokyo 113, Japan

P. Mustonen, Lipid Research Laboratory, Department of Medical Chemistry, University of Helsinki, Siltavuorenpenger 10, SF-00170 Helsinki, Finland

U. Nägele, Boehringer Mannheim GmbH, Werk Tutzing, Bahnhofstrasse 9-15, D-8132 Tutzing, Germany

K. Nowaczyk, University of Gdańsk, Institute of Experimental Physics, Gdańsk 80952, Poland

D. Oelkrug, Institut für Physikalische und Theoretische Chemie der Universität Tübingen, Auf der Morgenstelle 8, D-7400 Tübingen, Germany

F. Paltauf, Institut für Biochemie und Lebensmittelchemie, Technische Universität Graz, Petersgasse 12/II, A-8010 Graz, Austria

E. Prenner, Institut für Biochemie und Lebensmittelchemie, Technische Universität Graz, Petersgasse 12/II, A-8010 Graz, Austria

W. Rettig, I.N. Stranski Institute, Technische Universität Berlin, Strasse des 17. Juni 112, D-1000 Berlin 12, Germany

H. Riegler, Institut für Physikalische Chemie, Universität Mainz, Welder-Weg 11, D-6500 Mainz, Germany

R. Rigler, Karolinska Institute, Department of Medical Biophysics, Box 60400, S-10401 Stockholm, Sweden

W. N. Ross, Department of Physiology, New York Medical College, Valhalla, NY 10595, USA

T. Sakai, Department of Physiology, Tokyo Medical and Dental University, School of Medicine, 5-45, Yushima 1-chome, Bunkyo-ku, Tokyo 113, Japan

K. Sato, Department of Physiology, Tokyo Medical and Dental University, School of Medicine, 5-45, Yushima 1-chome, Bunkyo-ku, Tokyo 113, Japan

R. Schenk, Boehringer Mannheim GmbH, Werk Tutzing, Bahnhofstrasse 9-15, D-8132 Tutzing, Germany

W. Schmidt, Fachhochschule Aalen, Fachbereich Optoelektronik, Beethovenstrasse 1, D-7080 Aalen, Germany

H. Schneckenburger, Fachhochschule Aalen, Fachbereich Optoelektronik, Beethovenstrasse 1, D-7080 Aalen, Germany

S. G. Schulman, College of Pharmacy, University of Florida, Gainsville, FL 32610-0485, USA

N. J. Seare, Department of Chemistry, Loughborough University of Technology, Loughborough, Leicestershire LE11 3TU, UK

J. Slavik, Institute of Physiology, Czechoslovak Academy of Sciences, Videňská 1083, CS-142 20 Prague, Czechoslovakia

A. Sommer, Institut für Biochemie und Lebensmittelchemie, Technische Universität Graz, Petersgasse 12/II, A-8010 Graz, Austria

S. Summerfield, Department of Chemistry, Loughborough University of Technology, Loughborough, Leicestershire LE11 3TU, UK

H. Szmacinski, University of Maryland, School of Medicine, Center for Fluorescence Spectroscopy, Department of Biological Chemistry, 660 West Redwood Street, Baltimore, MD 21201, USA

W. Trettnak, Institute for Optical Sensors, Joanneum Research, Steyrergasse 17, A-8010 Graz, Austria

S. Uhl, Institut für Physikalische und Theoretische Chemie der Universität Tübingen, Auf der Morgenstelle 8, D-7400 Tübingen, Germany

M. Valcárcel, Department of Analytical Chemistry, Faculty of Sciences, University of Córdoba, E-14004 Córdoba, Spain

A. J. W. G. Visser, Department of Biochemistry, Agricultural University, Dreijenlaan 3, NL-6703 HA Wageningen, The Netherlands

B. Vogt, Boehringer Mannheim GmbH, Werk Tutzing, Bahnhofstrasse 9-15, D-8132 Tutzing, Germany

E. R. Weiner, U.S. Geological Survey, P.O. Box 25046 MS 424, Denver Federal Center, Lakewood, CO 80225, USA

J. Widengren, Karolinska Institute, Department of Medical Biophysics, Box 60400, S-10401 Stockholm, Sweden

O. S. Wolfbeis, Analytical Division, Institute of Organic Chemistry, Karl-Franzens University, A-8010 Graz, Austria

Part 1. New Methods in Fluorescence Spectroscopy

1. Fluorescence Spectroscopy: Where We Are and Where We're Going

Steven G. Schulman

College of Pharmacy, University of Florida, Gainesville, FL 32610-0485, USA

1 Introduction

Fluorescence, phosphorescence and chemiluminescence spectroscopies have established themselves as safe and ultrasensitive analytical techniques. However, the luminescence spectroscopies of today have moved well beyond the simple lamp, monochromator, sample compartment and detector arrangements of three or four decades ago, to very sophisticated electronic and optical instruments and some very elegant chemical and biochemical techniques and have drawn heavily on some of the most sophisticated physical and chemical theory for their current status. In this paper will be described some of the instrumentation and chemistry that have contributed to making luminescence spectroscopy one of the most powerful weapons in the chemist's and biochemist's arsenal and some of those approaches that will lead the way of this subject into the near future.

Fluorescence spectroscopy and the closely related areas of phosphorescence spectroscopy and chemiluminescence spectrometry have become firmly established and widely employed techniques of analytical chemistry. Fluorimetry and chemiluminescence spectrometry are now routinely used in the detection, quantitation, identification and characterization of structure and function of inorganic and organic compounds and of biological structures and processes. Phosphorimetry, although somewhat less popular, has the potential to be as useful as the other two areas if problems associated with sampling can be overcome.

Not so long ago it appeared that luminescence spectroscopy had reached an impasse as an area of active research in analytical chemistry. It was recognized that luminescence spectra lacked the information content of, say, mass spectra or nuclear magnetic resonance spectra and their intensities, as regards quantitation, were dramatically influenced by contaminants in solution. Coincident spectra from analytes with nearly identical fluorophores (e.g. catecholamines) made multicomponent analysis nearly impossible.

Chemiluminescence analysis was based on very few luminescent molecules and effective derivatization of analytes with these luminophores was a major problem. Phosphorescence spectroscopy required the use of very low temperatures. The use of narrow bore cylindrical quartz tubes (square cells crack at the joints in the freeze-thaw cycle) as cells for phosphorimetry entailed major difficulties with cleaning and filling as well as with repositioning for reproducible measurements. Fluorescent probes, used by bioanalytical chemists to examine the structures and functions of biological macromolecules and cellular organelles gave only the vaguest information about the "polarities" of binding sites and interchromophoric distances. Fluorescence microscopy was a purely qualitative tool.

In the past twenty years or so, much of this has changed. Thanks to recent advances in instrumentation and in new techniques, fluorescence and chemiluminescence spectroscopies can be routinely applied to real analytical problems. In liquid chromatography and immunoassay, fluorescence introduces ultrasensitive detection to extremely selective separation methods. Chemiluminescence spectrometry can also often be coupled with the selectivity of the immune response to allow sensitive quantitation, often without chemical or physical separation of the analyte from an extremely "dirty" matrix. With modern spectroscopic instrumentation, several analytes having nearly identical fluorescence spectra can often be determined simultaneously by taking advantage of differences in their fluorescent decay times or differences between the phase angles between their excitation and emission spectra rather than between their fluorescence spectral band intensities. Novel fluorescent probes targeted to particular organelles or chemical species found only in particular organelles have provided specific insight into some of the structure and function of proteins, nucleic acids and membranes. Phosphorimetry, on the other hand seems to be advancing at a somewhat slower pace than the other branches of luminescence spectroscopy.

In this paper, several modern developments in and applications of luminescence spectroscopy will be considered. Some of these have already been briefly mentioned. In addition this paper will consider some of the directions this author believes are likely to come under heavy investigation and development in the not-too-distant future.

2 The Marriage of Luminescence Spectroscopy With Separation Methods

The lack of specificity in luminescence analysis is based on the breadth and featurelessness of luminescence spectra and the similarities in spectral bandpes and spectral positions of the luminescence spectra of many compounds. to solve this problem include alteration of the bandform to introduce specificity and separation of interfering constituents from each other prior to quantitation by luminescence. In the latter approach, it was only natural that the simplicity and rapid development of high performance liquid chromatography and related separation methods would be coupled to luminescence spectroscopy in order to take advantage of the sensitivity of the latter and the specificity of the former. Fluorescence spectroscopy using conventional flow cells and monochromated or filtered arc-lamp excitation sources has been applied to detection in HPLC for over a decade now [1] and a wide variety of commercial detectors is currently on the market. The laser as an excitation source has been highly touted [2] but has, in this investigator's opinion, been somewhat disappointing. The laser, although many orders of magnitude more powerful as an excitation source than an arc-lamp, has, so far, produced detectability only one to two orders of magnitude better than that achieved with arc-lamps. This is a result of high background luminescence produced by photo and thermal decomposition of solution components induced by the large energy output of the laser. Thermal lensing effects also produce spectroscopic complication. Much more work needs to be done before the high cost of the ion-dye laser combination, as an excitation source, can be justified. On the other hand, the high degree of collimation of the laser beam has made it an excellent excitation source with the microbore cells and tubing used in capillary-zone-electrophoresis and microbore HPLC [3]. A typical argon-ion or dye laser beam can be focused into a spot no more than a few microns in diameter. With laser-excitation and fluorescence detection in capillary zone electrophoresis, analyte concentrations below one attomole (at nanomolar concentration) have been detected. Chemiluminescence has, in the past decade been coupled to HPLC, with promising results [4]. However, the multitude of reagents required to generate chemiluminescence, especially when energy transfer from the primary luminophore (the chemiluminescing molecule) to a spectroscopically superior acceptor (the secondary luminophore) is required, necessitates the use of many pumping systems and a very complex, if not unwieldy, chromatographic system. Moreover, although HPLC (and flow injection analysis) provide chemiluminescence spectrometry with a degree of specificity not inherent to the spectroscopic technique alone, the dynamic characteristics of the flowing separation systems reduce the residence time of the luminescing system in the detector. It is precisely the feature of ability to observe and integrate the chemiluminescence signal over the entire course of the lumigenic reaction that gives chemiluminescence its great sensitivity. Consequently, the state of the art being what it currently is, the features of the modern separation methods which account for great specificity also decrease the sensitivity of chemiluminescence analysis. Much work remains to be done in this area.

Detection by phosphorimetry in HPLC has been considered and its development to date, has been pioneered largely by the group at the Free University of Amsterdam [5]. To date, much of the work in this area has centered about using biacetyl (2,3-butanedione) as a phosphorophore. This molecule is one of very few that phosphoresce in fluid solution at ambient temperatures although rigorous deoxygenation of the solvent is necessary to observe biacetyl phosphorescence. Attempts to use directly excited phosphorescence and sensitized phosphorescence as detection methods have, so far, not proven practical. Even phosphorescence from phosphorophores entrapped in micelles or cyclodextrins, to protect them from dissolved oxygen and, hence, nonradiative deactivation has proven rather unfruitful. Phosphorescence efficiencies at room temperature tend to be rather low. Perhaps a bit more promising is the prospect of using a phosphorescent solute to provide a background and allowing the analyte to quench the background phosphorescence in the detector cell [5]. Because organic molecules that phosphoresce in fluid solution at ambient temperatures have lifetimes of the order of milliseconds (fluorescent molecules have lifetimes of the order of nanoseconds), dynamic quenching of phosphorescence can be about a million fold more sensitive than that of fluorescence – so, in principle, while it is impossible to detect quenching of fluorescence at concentrations lower than say 10^{-4} mol l^{-1}, quenching of phosphorescence can be detected at concentrations greater than 10^{-10} mol l^{-1}. Detection in HPLC, by phosphorimetry, still has a long way to go before it becomes suitable for routine analysis and it may well represent a major area of research in the future.

3 Improvement of Selectivity Based Upon Narrow-Band Excitation

These special techniques, depending largely on the narrow bandwidth of the laser as an excitation source, are based upon the uniqueness of the fine structure that appears in fluorescence spectra when inhomogeneous band broadening, especially that due to multiplicity of solvation sites of the fluorophore, is reduced with lowering of vibrational and rotational temperatures.

3.1 Fluorescence Line Narrowing (FLN) Spectroscopy

In fluorescence line narrowing spectroscopy [6, 7] a solution of a fluorophore is frozen at 4–10 K, and a dye laser is then used to excite part of the excitation spectrum corresponding to one of many solvent sites of the molecule, to get narrow line spectra. Freezing reduces the large number (continuum) of subtransitions in the excitation spectrum and makes the remaining ones appear discrete. Excitation must occur within 100–300 cm^{-1} of the 0–0 transition to obtain site selection. More than one site is excited when 0–1 excitation occurs. With narrow-line excitation, instead of relatively broad emission bands, narrow lines with structure called phonon wings are observed. Phonon wings are formed as a consequence of the coupling of solvent lattice vibrations with the electronic transitions of the fluorophore. The zero phonon line is the result of the purely electronic transition. In most solvents, at temperatures greater than 10 K, the phonon wings are large compared to the zero phonon line. As the temperature increases the phonon wings increase in size while the zero phonon line becomes smaller, until a broad band appears. Generally, above 30–50 K no narrow lines are obtained.

The laser is the ideal narrow-line excitation source. Nonlasing plasma lines are removed using a monochromator. Solvent Raman lines that may appear in the spectra are easily identified by comparisons to solvent blanks or spectra taken at higher temperatures. The best detection is obtained with a pulsed source and gated detector since Raman scatter is eliminated. If a continuous-wave laser is used stray light may be minimized with a double monochromator.

Solvents that form clear glasses at low temperatures are preferred as less scatter of exciting and emitted light occurs with these. 3-Methylpentane, 2-methyltetrahydrofuran, ethanol, methanol, ethanol-isopentane-diethyl ether mixtures, and mixtures of glycerol-water-ethanol are the solvents most often used. Cracking of the frozen glasses must be avoided as scatter from cracked solvents distorts the measured spectrum.

A helium bath cryostat gives the lowest temperatures and cools efficiently. Bath cryostats with variable temperature controls and conductance cryostats have also been used.

A large number of compounds has been analyzed using fluorescence line narrowing spectroscopy. Because of the site-selectivity inherent in the method, it has been successfully used to identify loci of damage in DNA produced by addition reactions with carcinogens derived from polycyclic aromatic hydrocarbons [8].

3.2 Shpol'skii Spectroscopy

Narrow spectral lines in the low temperature emission spectra of some aromatic molecules and other compounds in frozen n-alkanes were observed by Shpol'skii in 1952 [9]. The phenomenon is a matrix-induced effect: a polycrystalline solid matrix is formed by freezing the n-alkanes, producing a more homogeneous environment for the guest molecules, so that narrow spectral bands are observed. Typical sub-bandwidths are $1-10 \text{ cm}^{-1}$, although a temperature dependence of the 0–0 transition region of the emission spectrum is evident. The electron-phonon coupling is weak for guest-host systems so that well-resolved spectra may be seen at temperatures as high as 77 K. The basic requirement for Shpol'skii spectroscopy is the solubility of the sample in n-alkanes, which limits the scope of the method's use; however, this problem may be circumvented by dissolution of the sample in another solvent and subsequent dilution with an appropriate n-alkane [10].

The samples are cooled very quickly by immersion in liquid nitrogen or helium using Dewar flasks, or are cooled in a closed-cycle or continuous flow helium cryostat. A closed-cycle refrigerator has the advantage of achieving minimum temperatures of 10–15 K, by comparison with the 77 K obtainable in a Dewar flask of liquid nitrogen. Improved spectral resolution is probably brought about by the lower temperature.

The samples are excited by either a strong continuum source such as a xenon arc in combination with a high-resolution monochromator, an X-ray source [11, 12] or by a laser [13]. The matrix is opaque and strongly scatters light; the scattered light may be removed using filters since the sample need not be excited near the 0–0 transition. The number of compounds that show highly resolved emission spectra due to the Shpol'skii effect is rather limited and therefore, so is its analytical utility at present.

3.3 Matrix Isolation Fluorescence Spectroscopy

In this approach [14–16] the fluorophore is vaporized, mixed with a large excess of an inert gas (the matrix gas), and deposited on a cryogenic surface held at 15 K or less. Sample preparation requires the use of a closed-cycle cryostat. The inert gases Ne, Ar, and Xe, and N_2, n-perfluoroalkanes, and n-alkanes, have been used as matrix gases. The matrix gas is chosen so as to minimize the formation of solute aggregates in the matrix. Because of the low temperature of the sample the probability of diffusion of the solute molecules or rearrangement of the matrix lattice is very low. Chemical reactions of the solute with the matrix are not expected to occur because of the inert nature of the matrix. The technique is used to reduce interaction between analyte molecules and to give each an inert environment.

When organic compounds in nitrogen or argon matrices are excited using conventional lamp sources the observed spectral resolution is slightly better than that obtained in glassy frozen solutions, but the linewidths are wider than those obtained using Shpol'skii matrices, indicating that inhomogeneous line broadening is a problem. The use of lasers as excitation sources in matrix isolation fluorescence spectroscopy reduces the inhomogeneous line broadening. Narrower, Shpol'skii-like spectra may be obtained even when using conventional sources when matrix gases such as the n-perfluoroalkanes or n-alkanes are used. This procedure can be readily applied to samples which are thermally stable and are easily vaporized, but is not well suited to compounds that decompose when vaporized. Sampling, thus, is the area where the most immediate advances are needed, if the method is ever to be widely employed. Deposition of the solute is usually achieved through a slow, continuous, spray-on process although newer techniques of pulsed deposition are available. Pulsed deposition has the advantage of being faster than spray-on deposition, reducing the chances of impurities being introduced through leaks in the vacuum system. The leading proponent of matrix isolation fluorescence spectroscopy in analytical chemistry has been Wehry and his group at the University of Tennessee.

3.4 Supersonic Jet Spectroscopy

In this technique [17–22] high spectral resolution is brought about by Joule–Thompson Cooling in the gas phase and it stands in contrast to the three previous methods which are based upon solid-phase spectroscopic measurements.

A large drop in the internal energy of gas molecules is achieved when a gas is expanded through a narrow orifice from a high-pressure region. To escape, the molecules must acquire a large-velocity component in the axial direction. As the molecules leave the reservoir they collide with each other, leading to an averaging of the molecular velocities and a narrowing of the velocity distribution of the gas. Subsequently, a lower translational temperature is achieved.

An increase of the mass flow velocity, u, and a decrease of the local speed of sound, a, occurs during expansion. The flow conditions are described by the Mach number, M.

$$M = u/a$$

at some point in the expansion, M becomes greater than 1. Hence the flow of molecules described is known as a *supersonic jet*. Cooling in the supersonic jet leads to lower vibrational and rotational temperatures and, hence, superior spectral resolution. However, the fluorophores studied must obviously be relatively volatile and, in general, only laser-induced fluorescence emission and excitation spectroscopy can be used in conventional supersonic jets because of the small number of molecules that may be probed.

Experimentally, supersonic expansion spectroscopy is limited by the requisite of sufficient pumping capacity to handle the gas flow through the nozzle, which is mainly determined by the orifice assembly. There is lower sensitivity with a decrease in pressure due to an increase in dilution. The system is quite expensive, complicated and selective (e.g., in order to set up the exciting laser's wavelength, one has to know what the sample is). This approach has possibilities but is undoubtedly a long way from routine analytical use.

4 Fiber Optic Sensors – Non-Destructive Spectroscopy

Fiber Optic Sensors or "Optodes" are to spectroscopy what electrodes are to electrochemistry; non-destructive, regenerable analytical probes. In a fiber optic sensor, light from a source travels along an optically-conducting glass or plastic fiber to its end where absorption, reflection, scattering of light or fluorescence occurs. Fluorescent light travels up a second fiber and is collected by a detector system. Fluorescence sensors comprise the largest group of fiber optic sensors; fluorescence sensors based on principles of direct fluorescence measurement [23, 24], fluorescence quenching [25, 26] or competitive-binding [27] are available.

Instrumentation essentially consists of a light source, optical filters (if necessary), the fiber optic, a sensing zone, and a detector. Lasers, xenon lamps, hydrogen-, deuterium-, mercury-, and halogen lamps, and light emitting diodes (LEDs) have been used as excitation sources. Lasers are high intensity sources with a narrow bandwidth and are especially suited for remote sensing where light losses are large, but they are quite expensive and require a heavy power supply. The most promising light sources are the LEDs; these run at a low voltage and current, are small, have long lifespans, and do not generate heat, but have the disadvantage of working only from the IR region down to the blue. The material of the optical fiber determines the wavelength used. Fused silica, glass, and plastic fibers have all been used. Silica can be used for the ultraviolet range down to 220 nm but the fibers are expensive. Glass is suitable for use in the visible region and is reasonably priced. Plastic fibers are the cheapest but are limited to use above 450 nm. Measurements may be processed directly using analog and digital circuits or a microprocessor. Photomultipliers, photodiodes, photoconductive cells, and photovoltaic light detectors have all been used as detectors.

Fiber optic sensors are cheaper, more rugged, and smaller than electrodes; in the future we may see the former replacing the latter in various areas of analytical and clinical chemistry. Fields of application include biomedical telemetry, groundwater and pollution monitoring, process control, remote spectroscopy in high risk areas with radioactive, explosive, biological, or other hazards, titrimetry, and biosensing. The field has been an ever-expanding one due largely to the efforts of Seitz [28, 29] in the United States and Wolfbeis and his group [30] in Austria.

5 Semiconductor Lasers

Semiconductor lasers (laser diodes) have several outstanding advantages over gas and dye lasers and arc-lamps as excitation sources [31]. They are more powerful and more coherent light sources than arc-lamps. They also have the advantage of producing a polarized beam which can be useful for certain applications. They are smaller, more compact and much less expensive than other kinds of lasers (albeit less powerful) and the fact that all semiconductor lasers emit in the far red or infrared means that less energy can be deposited in the sample so that considerably less thermal decomposition can occur than with conventional ion, excimer and dye lasers. The fact that emission from semi-conductor lasers is confined to the far red and infrared is a "mixed blessing". On the one hand, the fact that so few substances demonstrate electronic absorption in that region of the electronic spectrum means that excitation of semiconductor laser light will be very selective. For example, the oft-interfering luminescences of tryptophan and bilirubin from serum samples will not be problematic with semiconductor laser excitation because these substances are not excited by red light. On the other hand, since so few substances absorb red light, a new generation of fluorescent probes and labels is required for the labeling of analytes to be detected subsequent to chromatographic or immunochemical analysis.

Several of the polymethine dyes absorb in the red, far red and near infrared and fluoresce efficiently as well in this spectroscopic region. Functional derivatives of these may provide excellent fluorescent labels for semiconductor laser excitation. Additionally the groups led by Akiyama in Nagasaki, Wolfbeis in Graz, Miller in Loughborough and (the late) Ishibashi in Sendai are actively involved in the synthesis and development of large polyunsaturated dye molecules which should be amenable to excitation and, hopefully, luminescence in the near infrared [32, 33].

Other areas of luminescence spectroscopy that are in use today and are areas of rapid growth are time and phase resolved luminescence spectroscopy

[34] and the closely related area of frequency domain spectroscopy [35], total luminescence spectroscopy [36], immunoassay [37, 38], especially with exotic labels such as rare earths and chemiluminophores [39, 40] and flow cytometry [41]. These subjects have been reviewed recently and space limitations prevent extensive discussions of these areas here. It is certain that luminescence spectroscopy in all its forms is a viable and growing area; one which will be preeminent in the basic and applied sciences for many years to come

Acknowledgement. The author wishes to express his gratitude to Ms. Patricia Khan and Mrs. Luz E. Jimenez for technical assistance with the preparation of this manuscript.

6 References

1. Hulshoff A, Lingeman H (1985) Fluorescence detection in chromatography. In: Schulman SG (ed) Molecular luminescence spectroscopy: Methods and applications, Part 1, Wiley-Interscience, New York, p 621
2. Van den Beld CMB, Lingeman H, Van Ringen GJ, Tjaden UR, van der Greef J (1988) Anal Chim Acta 205: 15
3. Cheng YF, Dovichi N (1989) Science 242: 62
4. Kobayashi K, Imai K (1980) Anal Chem 52: 424
5. Gooier C, Baumann RA, Velthorst NH (1987) Progr Anal Spectrosc 10: 573
6. Hofstraat JW et al. (1987) Anal Chim Acta 193: 193
7. Wehry EL, Mamantov G (1981) In Molecular fluorescence spectroscopy, vol. 4 (Wehry EL ed). Plenum Press, New York, p 193
8. Heisig V et al. (1984) Science 223: 289
9. Shpol'skii EV et al. (1952) Dokl Akad Nauk SSSR 87: 935
10. Chen WS et al. (1984) Can J Chem 62: 2264
11. Woo CS et al. (1978) Environ Sci Technol 12: 173
12. Woo CS et al. (1980) Anal Chem 52: 159
13. D'Silva AP, Fassel VA (1984) Anal Chem 56: 985A
14. Meyer B (1971) Low Temperature Spectroscopy, Elsevier, New York
15. Hallam HE (ed) (1973) Vibrational spectroscopy of trapped species. Wiley, London
16. Craddock S, Hinchcliffe AJ (1975) Matrix isolation: A Technique for the study of reactive inorganic species, Cambridge University Press, Cambridge
17. Smalley RE et al. (1977) Acc Chem Res 10: 139
18. Levy DH et al. (1977) In: Moore CB (ed) Chemical and biochemical applications of Lasers, vol 2 Academic Press, New York, p 1
19. Levy DH (1980) Ann Rev Phys Chem 31: 197
20. Levy DH (1981) Science 214: 263
21. Hayes JM, Small GJ (1983) Anal Chem 55: 565A
22. Johnston MV (1984) Trends Anal Chem 3: 58
23. Saari LA, Seitz WR (1982) Anal Chem 55: 667
24. Zhujun Z, Seitz WR (1986) Anal Chem 58: 220
25. Lubbers DW, Opitz N (1983) In: Proceedings of the International Meeting on Chemical Sensors, Fukuoka, Japan, Elsevier, Amsterdam, p 609
26. Peterson JI et al. (1984) Anal Chem 56: 62
27. Schultz JS et al. (1982) Diabetes Care 5: 245
28. Seitz WR (1984) Anal Chem 56: 16A
29. Seitz WR (1987) J Clin Lab Anal 1: 313
30. Wolfbeis OS (1991) Fiber optic chemical sensors and biosensors, CRC Press, Boca Raton, Fl., Vols. I & II
31. Kawabata Y, Imasaka T, Ishibashi N (1986) Talanta 33: 281
32. Akiyama S, Nakatsuji S, Nakashima K, Watanabe M, Nakazumi H (1988) J Chem Soc Perkin Trans I: 3155

33. Nakatsuji S, Nakazumi H, Fukuma H, Yahiro T, Nakshima K, Iyoda M, Akiyama S (1990) J Chem Soc Chem Commun, 489
34. Demas JN (1988) Time-resolved and phase-resolved emission spectroscopy. In: Schulman SG (ed) Molecular luminescence spectroscopy: Methods and applications, Part 2, Wiley-Interscience, New York, p 79
35. Lakowicz JR, Maliwal BP (1986) Biophys Chem 21: 61
36. Leiner MJP, Hubmann MR, Wolfbeis OS (1991) In: Baeyens WRG, DeKeukelaire D, Korkidis K (eds) Luminescence techniques in chemical and biochemical analysis, Dekker, New York, p 381
37. Bentz AP (1976) Anal Chem 48: 455A
38. Warner IM, Callis JB, Davidson ER, Christian GD (1976) Clin Chem 22: 1483
39. Schulman SG, Hochhaus G, Karnes HT (1991) Fluorescence Immunoassay. In: Baeyens WRG, DeKeukeleire D, Kordidis K (eds) Luminescence techniques in chemical and biochemical analysis, Dekker, New York, p 341
40. Jansen EHJM, Van Petaghem CH, Zomer G (1991) Chemiluminescence immunoassays in veterinary and food analysis. In: Baeyens WRG, DeKeukelaire D, Korkidis K (eds) Luminescence techniques in chemical and biochemical analysis, Dekker, New York, p 477
41. Muirhead KA, Horan PK, Poste G (1985) Biotechnology 3: 337

2. Interactions and Kinetics of Single Molecules as Observed by Fluorescence Correlation Spectroscopy

Rudolf Rigler, Jerker Widengren and Ülo Mets

Department of Medical Biophysics, Box 60400, S-10401 Stockholm, Sweden

1 Introduction

Random fluctuations of the intensity of individual molecules excited to fluorescence by a stationary light source provide information on important molecular properties such as rotational motion [1–5], translational diffusion [6, 7], chemical kinetics [7–9] as well as the lifetime of the excited state [1, 2, 10].

The idea of this analysis is to observe the intensity fluctuations of molecules excited to fluorescence in a tiny volume element. These variations are due to fluctuations in the number of molecules when entering or leaving the volume element (translation) or which are caused by changes in the orientational excitation probability (rotation), the quantum yield (due to interactions and conformational transitions) as well as by depletion and repopulation of the ground state (life time). Contrary to relaxation experiments which depend on external perturbations, the systems' thermodynamic fluctuations themselves are observed.

The signal analysis is based on calculating the correlation function of the fluorescence intensity fluctuations as shown by Elson and Magde [6, 7] for the case of translational diffusion and by Ehrenberg and Rigler [1, 2] for the case of rotational motion and its coupling to the decay of the excited states. The success of this analysis, as pointed out by Koppel [11], rests on the ability to detect a large enough photon flux per molecule. Usually photo-chemical destruction of the molecular species under observation has been a limiting factor for fluorescence correlation analysis.

We have recently been able to improve the detected photon flux/molecule by orders of magnitude without measurable photodestruction by employing extremely small open volume elements (10^{-15} l and below) in combination with confocal epi-illumination [12, 13]. The characteristic diffusion times are thereby reduced substantially in relation to photobleaching times. High aperture confocal optics in combination with solid state detectors yield a maximum detection efficiency and background discrimination. In a typical experiment the autocorrelation function of a few molecules per volume element is obtained in seconds or less with a sample of a few µl.

2 The Experimental Set up

The set up consists of a Spectra Physics Model 165 Argon laser which is focussed on the image plane of a Zeiss Plan-Neofluar $40\times$ NA 0.9 or a Zeiss

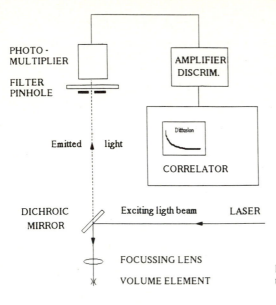

Fig. 2.1. Set up for fluorescence correlation measurements

Plan-Neofluar $63\times$ NA 1.2 objective which in turn refocusses the laser beam in an epi illumination arrangement (Fig. 2.1). The emission is detected through a pinhole (30–90 μm) conjugated with the image plane of the volume element and detected by an EG&G avalanche photodiode SPCM-100 or a Hamamatsu R929 photomultiplier operated in the photon counting mode. A cutoff filter (Schott KV 550) is used to discriminate against light scattered from the primary excitation source. Depending on the magnification and pin holes used the volume elements vary between 0.2 fl ($63\times$, 30 μm pinhole) and 12 fl ($40\times$, 90 μm pinhole). Neutral glass filters were used to attenuate the laser beam to give an intensity 0.5 mW and lower at the sample plane. In the case of solutions a hanging drop was placed under the microscope objective while investigations on membranes were done with a cover slip and glycerol immersion.

The detection efficiency of the set up may be demonstrated by the fact that for the smallest volume element used the measured photon counting rate for one molecule Rhodamine 6G in water was 62000 cps and that of the background 1300 cps yielding a signal to background ratio of 47 with 0.5 mW excitation at 514.5 nm.

The correlation functions were calculated in real time by a 4×4 bit Langley–Ford 1096 correlator. The data analysis was performed using non linear least square parametrization [14] with the statistical criteria given by Koppel [11] for calculating the normalized mean square deviation between data and model (X^2).

3 Theoretical Background

The intensity of molecules excited to fluorescence in solution varies due to spontaneous fluctuation in the particle number in an open volume element and gives rise to an average photon flux $\langle i \rangle$ and its fluctuating part $\delta i(t)$. For the analysis of characteristic time constants the autocorrelation function $C(t)$ is calculated:

$$C(t) = \langle i(t)i(0) \rangle = \lim_{T \to \infty} \frac{1}{2T} \int_{-T}^{T} \langle i(t + t_1)i(t_1)dt_1 \rangle, \tag{1}$$

$C(t)$ has a constant part $\langle i \rangle^2$ and a time dependent part so that

$$C(t) = \langle i \rangle^2 + \langle \delta i(t)\delta i(0) \rangle. \tag{2}$$

Information about the kinetics in the process under investigation is contained in the term $\langle \delta i(t)\delta i(0) \rangle$. General considerations of the statistics of Poissonian processes lead to the conclusion that the relative amplitude of $C(t)$ is related to the absolute number of particles in the volume element N:

$$\lim_{t \to 0} \langle \delta i(t)\delta i(0) \rangle / \langle i \rangle^2 = \frac{1}{N}, \tag{3}$$

with $N = cV$, c being the particle concentration and V the size of the volume element.

A general solution of the autocorrelation function has been given by solving the Fokker-Planck equation for rotational and translational diffusion including excited state – ground state transitions [1, 2]. For the situation that the fluorescence decay (τ) rotational motion (ρ) and translational diffusion (τ_{diff}) are well separated in time

$$\tau \ll \rho \ll \tau_{\text{diff}},$$

the normalized autocorrelation function

$$G(t) = C(t)/\langle i \rangle^2 \text{ takes the form:}$$

$$G(t) = 1 + \frac{1}{N} \left\{ \frac{1}{1 + 4Dt/w^2} + \frac{4}{5}e^{-t/\rho} - \frac{9}{5}e^{-t/\tau} \right\}, \tag{4}$$

with
τ = fluorescence lifetime
ρ = rotational relaxation time
$D = \dfrac{w^2}{4\tau_{\text{diff}}}$ = translational diffusion constant
w^2 = radius of the volume element (e^{-2} point of the Gaussian beam).

A schematic view of rotation and translation in the volume element is given in Fig. 2.2.

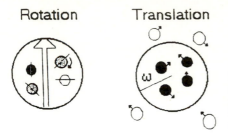

Fig. 2.2. Cross section of volume element with molecules emitting photons at different rates (black, grey, white) due to varying orientation relative to the polarization vector (rotation) and varying position in the Gaussian beam (translation)

Contrary to the case of the orientational randomization after molecules have been elevated to the excited state by a pulse of polarized light [15, 16], rotational diffusion as observed by FCS is not linked to the lifetime of the excited state. This fact renders anisotropy decay insensitive for the measurement of slow rotations as observed e.g. in membranes. The existence of anticorrelations due to the negative sign of the excited states' decay which we have predicted [1] has been demonstrated experimentally by Kask et al. [10].

The correlation function for translational diffusion

$$G(t) = 1 + \frac{1}{N}\left(\frac{1}{1 + 4Dt/w^2}\right), \tag{5}$$

relates to a volume element with Gaussian intensity distribution in the x, y plane and infinite dimension in the z-direction. It can be visualized as a limiting case of a finite volume element with a Gaussian intensity distribution in three dimensions [3, 12, 13]

$$G(t) = 1 + \frac{1}{N}\frac{1}{1 + 4Dt/w_{xy}^2}\left\{\frac{1}{1 + 4Dt/w_z^2}\right\}^{1/2}. \tag{6}$$

The coupling between translational diffusion and chemical reactions has been treated extensively by Elson and Magde and we refer to the original publications [6, 7]. Two limiting cases are of interest:

(a) the chemical relaxation times are much smaller than the characteristic diffusion times, i.e. the chemical exchange is equilibrated during diffusion through the volume element: $\tau_{chem} \ll \tau_{diff} = w^2/4D$. We obtain for

$$G(t) = 1 + \frac{1}{N}\left\{\frac{1}{1 + 4\langle D\rangle t/w^2}\right\}, \tag{7}$$

with

$$\langle D\rangle = xD_{free} + yD_{bound},$$

where $\langle D\rangle$ is the weighted average of the diffusion of the free and the bound ligand with x and y denoting the fraction of free and bound ligand.

(b) the chemical relaxation times are much larger than the characteristic diffusion times and the chemical exchange does not take place on the time scale of diffusion: $\tau_{\text{chem}} \gg \tau_{\text{dif}}$:. For this case we obtain:

$$G(t) = 1 + \frac{1}{N} \left\{ \frac{x}{1 + 4D_{\text{free}}t/w^2} + \frac{y}{1 + 4D_{\text{bound}}t/w^2} \right\}. \tag{8}$$

Both situations are of interest since they allow the analysis of molecular interactions by determining the chemical equilibrium distribution from the knowledge of the diffusion constants of the free and bound ligand.

4 Results

4.1 Translational Diffusion

One of the improvements of the FCS technique lies in the very small open volume elements used permitting rapid exchange of the molecular species. In order to determine the volume element we have evaluated the diffusion constant

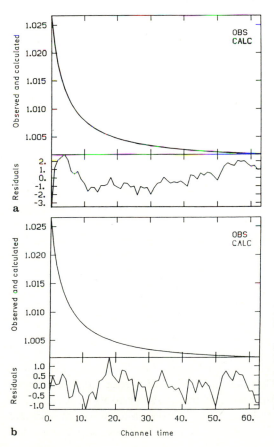

Fig. 2.3a, b. $G(t)$ of translational diffusion Rhodamin 6G 3×10^{-8} M, $V = 1.5$ fl, channel time 0.1 ms. **a** evaluation of N and τ_{diff} with 2-D analysis $(w_1/w_2 = 0)$ $N = 28$, $\tau = 0.12$ ms, $X^2 = 1.74$; **b** evaluation of N and τ_{diff} with 3-D analysis $N = 29$, $\tau = 0.13$ ms, $X^2 = 0.301$. Lower curve in panel residuals of least square fits

according to a 2-dimensional analysis (Eq. 5) as well as to a 3-dimensional analysis (Eq. 6). As is evident from Fig. 2.3a and Fig. 2.3b a considerable improvement of the fit shown by the reduced X^2 value and the residuals is observed as reported before [12] when the 3-dimensional model is applied. The volume element observed under this condition (40× magnification and 50 μm pinhole) is determined by its half axes $w_1 = w_{xy}$ equal to 0.4 μm and $w_2 = w_z$ equal to 1.6 μm yielding a total volume of 1.5 fl. With a concentration of 3×10^{-8} M Rhodamin 6G 29 molecules are observed in the volume element which enter and leave with a diffusion time $\tau_{\text{diff}} = w_1^2/4D$ of 134 μs. As seen from the reduced residuals and X^2 values a considerable improvement in the fit of the correlation function is observed with a 3D volume element as opposed to a 2D one leading to an increase in both τ_{diff} and N. The error in τ is about 7% and is dependent on the size of the volume element [13]. If diffusion occurs in a plane (2-dimensional diffusion) the results of 2D and 3D analysis should be indistinguishable.

4.2 Rotational Diffusion

FCS anisotropy measurements differ in one important point from the conventional time resolved analysis of rotational motion using pulsed excitation. In the FCS the rotation of the ground state is probed via its correlation with the excited state [1, 2] contrary to the fluorescence anisotropy decay where the rotation of the excited state is observed. An important feature of FCS is that the fluorescence lifetime is uncoupled from the rotational relaxation time and even slow rotations can be observed using rapidly decaying fluorescence probes. In Fig. 2.4a and 2.4b we demonstrate as example the rotational and translational diffusion of the acetylcholin receptor (Mw = 290000) which was isolated in E. Neumann's laboratory and which carries a rhodamin labelled α-bungarotoxin molecule, a highly affine inhibitor. Values of 1.6×10^{-5} s and 1.4×10^{-2} s were obtained from the analysis for the rotational relaxation and diffusion times respectively and indicate that the receptor exists in an aggregated form under the conditions used (Fig. 2.4.) Since the diffusion time is dependent on the radius of the volume element w but the rotational relaxation time is not, the effect of decreasing the volume element by increasing the magnification (63×) was tested. As expected, the diffusion time decreased with the decreased beam radius but not the rotational relaxation time (not shown). Both processes contribute by about equal amplitude to the correlation function as predicted from Eq. (4). The lifetime of the label tetramethyl rhodamin of 4 ns as determined in this laboratory is almost four orders of magnitude faster than the rotation time and fluorescence anisotropy decay could not have been used for this analysis. This example may demonstrate the efficiency of FCS in analyzing molecular transport properties. In particular membrane systems will be of considerable interest in order to analyze the coupling of rotation and diffusion of various membrane components [17].

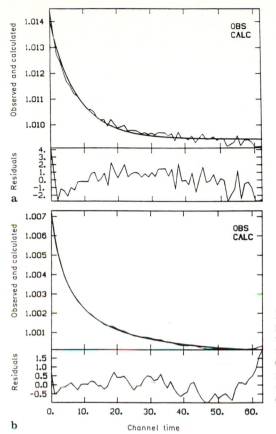

Fig. 2.4a, b. $G(t)$ of the acetylcholine-receptor from *Torpedo californica* labelled with tetramethylrhodamin-α-bungaro-toxin, in 10 mM Hepes, 0.1 M NaCl, 0.5% CHAPS with 50% glycerol (v/v). **a** rotational diffusion, channel time 2 µs, $\tau_{rot} = 16$ µs; **b** translational diffusion, channel time 3 ms $\tau_{diff} = 14.2$ ms

4.3 Molecular Interactions and Kinetics

The high sensitivity in recording correlation functions is shown by an example where translational diffusion of Rhodamin 6G was recorded for one molecule on average in the volume element (Fig. 2.5b). A correlation function was obtained in 20 ms which compared well with the correlation function recorded during longer time (Fig. 2.5a). With a diffusion time of about 100 µs one molecule has to cross the volume element ca 200 times in order to generate the correlation curve in Fig. 2.5b. This result stimulated us to further research in order to find out whether recording of the passage of a single individual molecule was within reach [18].

For the purpose of analyzing molecular interactions of high specificity characterized by high binding constants and low dissociation rates, the feasibility of recording fluorescence correlations in fractions of a second opens up a new and extremely sensitive method of analysis. As outlined by Eq. (8) the fraction of bound and free ligand can be followed if the diffusion constants or

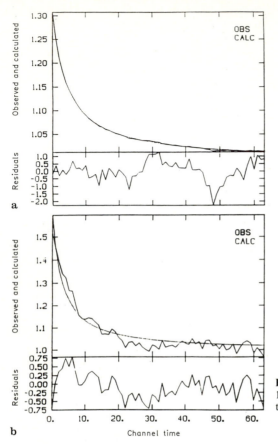

Fig. 2.5a, b. $G(t)$ of Rhodamin 6G, 1×10^{-9} M, channel time 0.1 ms. **a** run time 10 s; **b** run time 20 ms

diffusion times of free and bound ligand are known. Figures 2.6a and 2.6b show the result of a kinetic analysis where the association of rhodamin labelled α-bungarotoxin with the acetylcholine receptor was followed by a sequential recording of the autocorrelation function of the bungarotoxin diffusion after the addition of the acetylcholine receptor. The fraction unbound bungarotoxin which is decreasing in time was calculated. Contrary to what is expected from a diffusion controlled reaction even with the low concentrations used (α-bungaro-toxin 4×10^{-8} M, acetylcholine receptor 1×10^{-7} M) a surprisingly low association time was evaluated pointing to a very low recombination rate constant [19]. Corresponding results have been obtained using radioisotope labelled α-bungarotoxin [20]. Similarly the dissociation rates can be determined and thus the equilibrium constant.

4.4 Single Molecule Events

The extreme sensitivity of the method which has enabled registration of correlation functions of molecules in fM concentrations [12] as well as the

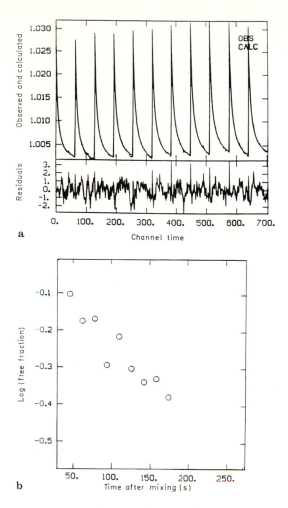

a

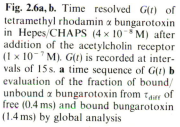

b

Fig. 2.6a, b. Time resolved $G(t)$ of tetramethyl rhodamin α bungarotoxin in Hepes/CHAPS $(4 \times 10^{-8} M)$ after addition of the acetylcholin receptor $(1 \times 10^{-7} M)$. $G(t)$ is recorded at intervals of 15 s. **a** time sequence of $G(t)$ **b** evaluation of the fraction of bound/unbound α bungarotoxin from τ_{diff} of free (0.4 ms) and bound bungarotoxin (1.4 ms) by global analysis

results shown in Fig. 2.5b have prompted us to investigate the present limit in our hands. From the particle number N, which in very low concentrations is very much smaller than unity (10^{-3} and below), the probability of single and double molecule transits can be estimated. With this information the occurrence of the passage of a single individual molecule through a volume element can be verified [18] and Fig. 2.7a shows the photon bursts of individual rhodamin 6G molecules in H_2O passing through a Gaussian volume element of 0.2 fl by Brownian motion with a diffusion time of ca 40 μs. When compared with similar reports [21] our data show an order of magnitude higher sensitivity.

5 Conclusions

The use of extremely small open volume elements in combination with confocal epi illumination has led to a large improvement of the S/N ratio in FCS

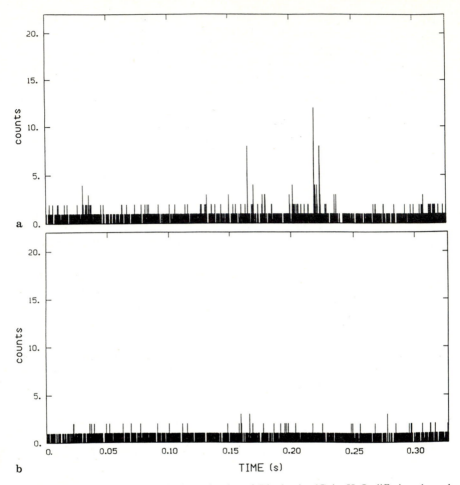

Fig. 2.7a, b. Photon bursts of single molecules of Rhodamin 6G in H$_2$O diffusing through a Gaussian volume of 0.24 fl. **a** multichannel scaling trace, channel time 40 µs; **b** background

experiments [12, 13]. This is due to the rapid and efficient exchange of molecules within the confined Gaussian volume of excitation thereby reducing photobleaching but also due to increased optical collection efficiency as well as detector efficiency.

A 3 dimensional volume element has to be used to describe accurately translational diffusion [12, 13]. For Rhodamin 6G with a diffusion constant of 2.8×10^{-6} cm^2 s^{-1} diffusion times of 40 µs are observed using the smallest $V = 0.24$ fl. We have been able to measure diffusion times up to 1 s which corresponds to diffusion constants between 10^{-9} to 10^{-10} cm^2 s^{-1}. Contrary to translational diffusion rotational motion is not dependent on the size of the volume element and can be separated from translation by the volume dependence of the correlation time. Coupling of rotational and translational diffusion can be easily followed in a FCS experiment. Its analysis is of particular importance for elucidating structure of membrane bound receptors and ligands.

Chemical kinetics due to biomolecular reactions and conformational transitions can be analyzed by FCS [6, 7] provided that diffusion times are of the same order or longer than chemical relaxation times. Short diffusion times due to small volume elements reduce the possibility to extract chemical reactions on a longer time scale on one hand, but. on the other hand they allow to follow reaction kinetics by watching the proportion of complexed ligands due to their different diffusion constants. Since low concentrations can be used even fast reactions can be observed by recording time resolved correlation functions which can be obtained in a fraction of a second. Equilibrium constants are available from titrations as well as association and dissociation rates. Extremely small amounts (1–100 fmol) are needed for the examination. The importance of this method for analyzing molecular interactions and kinetics in solution as well as on membranes is evident.

We have been able to record correlation functions with concentrations of 100 fM and lower [12], which means that the probability to find a molecule in the observed volume element is 10^{-4} ($N = 10^{-4}$). By optimizing the volume element a signal to background ratio of 40 or measuring 1 molecule was obtained [13]. We are subsequently able to record the transit of an individual molecule through the Gaussian volume element caused by Brownian motion. The photon bursts created by the transitions are much above the background. (Fig. 7) and the highest bursts relate to the transition through the Gaussian peak.

The possibility to observe single molecules in transit opens up ways to detect very rare events of chemical and biological significance. Also the chemical and biochemical analysis of compounds at the single molecular level should be possible.

Acknowledgment. We thank Mr. L. Wallerman for expert workshop assistance. The work was supported by funds from the K. & A. Wallenberg foundation, J. & L. Grönbergs foundation and the Technical Science Research Board.

6 References

1. Ehrenberg M, Rigler R (1974) Chemical Physics 4: 390
2. Ehrenberg M, Rigler R (1976) Quart Rev Biophys 9: 69
3. Aragon SR, Pecora R (1975) J Chem Phys 64: 1791
4. Rigler R, Graselli P, Ehrenberg M (1979) Phys Scripta 19: 486
5. Kask P, Piksarv P, Pooga M, Mets Ü, Lippmaa E (1989) Biophys J 55: 213
6. Elson EL, Magde D (1974) Biopolymers 13: 1
7. Magde D, Elson EL (1974) Biopolymers 13: 29
8. Icenogle RD, Elson EL (1983a) Biopolymers 22: 1919
9. Icenogle RD, Elson EL (1983b) Biopolymers 22: 1949
10. Kask P, Piksarv P, Mets Ü (1985) Eur Biophys J 12: 163

11. Koppel DE (1974) Phys Rev A10: 1938
12. Rigler R, Widengren J (1990) Bioscience 3: 180
13. Rigler R, Mets U, Widengren J, Kask P (1992) Submitted
14. Marquardt DW (1963) J Soc Ind Appl Math 11: 431
15. Ehrenberg M, Rigler R (1972) Chem Phys Lett 14: 539
16. Rigler R, Ehrenberg M (1973) Quart Rev Biophys 6: 139
17. Saffman PG, Delbrück M (1975) Proc Nat Acad Sci 72: 1975
18. Rigler R, Mets U (1992) Submitted
19. Rauer B (1992) Thesis, Univ of Bielefeld
20. Oswald RE, Changeux JP (1982) FEBS lett, 139: 225
21. Soper SA, Shera E, B. Martin JC, Jett JH, Hahn JH, Nutter HL, Keller RA (1991) Anal Chem 63: 432

3. Fast Optical Imaging Techniques

Craig D. MacKay

Astromed Ltd., Innovation Centre, Cambridge Science Park, Milton Road, Cambridge, CB4 4GS, UK

1 Introduction

The distinction between detectors for imaging and spectroscopy is becoming increasingly blurred. The use of two-dimensional detectors to allow the recording of many spectra simultaneously is rapidly becoming an essential part of many experimental set-ups where the requirement is to look at temporal or spatial spectral variations within one experiment. In many applications there is a need for good time resolution. Traditional systems are used that operate at the rate of 50 or 60 Hz commonly used by standard TV cameras. This paper will look at the limitations that are encountered in trying to take fluorescent images at much higher repetition (frame) rates.

2 High Speed Imaging Technologies

There are already well established methods of extremely high-speed imaging. These use streak or framing cameras that rely on a scanning image intensifier to switch an image onto a phosphor screen for a brief instant. The images are read from the phosphor screen by a conventional imaging system so only a small number of images may be recorded before the intensifier phosphor screen is fully covered. Only short time sequences may be covered but these methods allow the highest time resolutions currently available and these are in the sub-nanosecond region.

At the slow end of the range, systems which operate at TV frame rates are available. Often, however, their performance is barely adequate for the task of fluorescence imaging. We are always concerned not simply to take images rapidly but to ensure that the images are of a quality that will permit the measurements called for by the research programme in question. The quality of an imaging system is a combination of its spatial resolution, sensitivity and dynamic range. Let us consider each in turn.

2.1 Resolution

This is the measurement of the ability of a detector system to resolve details in the image in question. It is very important to distinguish this from the number of pixels in an image. Many detectors may be operated to give very large numbers of pixels, yet have poor intrinsic resolution.

2.2 Sensitivity

This is the measurement of the lowest light level that will give the required signal to noise ratio. It will be affected by the detector quantum efficiency and any internally generated noise sources such as dark current (the signal generated in the absence of any input light signal) and read-out noise (the noise from the system even if zero signal or dark noise is present). Good sensitivity means better signal to noise ratios. It can also mean that lower stimulating light levels may be used, lessening their effects on the sample under study.

2.3 Dynamic Range

This is the measure of the ratio of the brightest to faintest features that may be measured simultaneously in a single image. It also is a measure of the smallest signal level differences that may be measured (minimum detectable contrast). These numbers are somewhat dependent on resolution since signal averaging will sometimes improve contrast sensitivity and dynamic range. In fluorescence applications we are often concerned to see faint features against a bright background. Many conventional TV cameras will only achieve one percent contrast sensitivity under ideal illumination conditions while cooled, slow-scan CCD cameras can achieve extremely low contrast performance indeed (< 0.1 percent).

3 Imaging System Components

It is important to remember that a complete imaging system is made up of several components and any real system may be limited by any one of them. Of particular concern here is the detector system itself but often limitations in the control of illumination of the sample or the detector are important. Many detector systems require a shutter to be provided to cover the detector during read-out so as to avoid image smear. This can often be done by controlling the illuminating light source. Mechanical leaf shutters allow exposures down to a few milliseconds at up to a few tens or a hundred exposures per second. Rotating mirror shutters allow exposures down to the microsecond level though they are often difficult to incorporate in a fast efficient imaging system. Ferroelectric liquid crystal shutters can work at tens or hundreds of microseconds exposure time. However in white light they may give one or two percent transmission in their "closed" state and even at selected wavelengths, attenuation ratios of 1000:1 are difficult to achieve routinely. For many fast time resolution fluorescence applications it is easiest to control the light source with switched laser stimulation or xenon flash lamps.

Fluorescence measurements generally require that the detector can handle relatively fast photon arrival rates. If a particular feature is to be detected with a contrast of 1 percent of the background and a signal to noise ratio of 5 (a minimum for a two-dimensional imaging system) then there has to be $> 2.5 \times 10^5$ photons detected over the feature purely from photon statistical noise considerations. For many systems (such as photon counting systems) these photon rates, which may need to be recorded many times per second, which may simply lead to detector saturation and overload. It is also a great practical advantage for systems to be insensitive to substantial light overloads since mistakes are easy to make.

It is often tempting to specify an imaging system to have a large number of pixels. Many systems perform rather poorly, because their designers have been too greedy for pixels. Pixels all take time to read-out and the data they produce has to be transferred somewhere for storage and analysis. The first law of all image capture and image processing is to throw away all unwanted data as quickly as possible to make the data handling as manageable as possible. Computer systems are becoming more powerful every year but high speed data storage systems are very expensive and fill up remarkably quickly.

In this connection it is important to look carefully at the support hardware (usually a computer with software) needed to handle the imaging system and analyse the images generated. Many excellent detector sub systems are compromised by poor computer support facilities that cause the greatest frustration to the research worker.

4 Examples of Detector Systems

In this section the basic features of a variety of detector sub-systems are described.

4.1 Standard Silicon Photodiodes

These devices have good detection efficiency but are limited by their intrinsic noise sources to relatively high light levels (a typical noise level might be equivalent to 10^5 photons equivalent for a 1 kHz time resolution). They are available in arrays of a few hundred diodes (Centronics) with fairly coarse resolution (1–2 mm) each of which requires its own amplifier/signal processing chain. The use of these devices forces the user into a high light level mode of operation with the problems described earlier. However there are several groups who use them successfully.

4.2 Vidicon TV Cameras

These use magnetic or electrostatic deflection of a scanning beam to read-out a light sensitive target. Faster operation is difficult since the deflection coils require extremely high currents to achieve this and are generally limited by the time constants of the deflection circuitry. Dynamic range is generally poor and unless cooled are unable to achieve low-light level performance because of dark current build up. They are increasingly being replaced by charge-coupled device (CCD) cameras operated at a TV frame rate.

4.3 Charge-Coupled Devices (CCD)

There are many different operating methods for CCDs that can dramatically affect their performance. Standard CCD cameras may be purchased for a few hundred dollars that provide TV (50 or 60 frames per second) rate images at fairly good resolution (such as 640×480 pixels). They are difficult to push to faster read-out speeds because of slew-rate limitations intentionally built into the internal (on-chip) amplifier to minimise internally generated noise. Their intrinsic sensitivity is good since silicon is an efficient photodetector over a wide range of wavelength. At room temperatures they are usually limited by their internally generated dark current. This may be reduced to negligible levels by cooling. Then at video pixel rates (typically 8 MHz) the read-out noise limits sensitivity at levels of 150–200 electrons or above. Recently, however, systems have been developed that are able to achieve read-out noise levels of 15–30 electrons at this sort of pixel rate.

In addition, these systems allow operating modes where only a small part of the large CCD is read-out to give much faster frame repetition rates over a

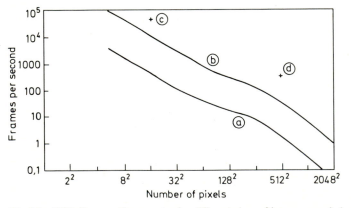

Fig. 3.1. CCD Camera Frame rates for different sizes of image. *a* cooled, slow-scan CCD camera system such as the Astromed 3200 system, *b* cooled programmable video-rate CCD camera system, *c* Thomson 14×14 pixel CCD, *d* the EEV CCD13 8 parallel output 512×512 pixel CCD

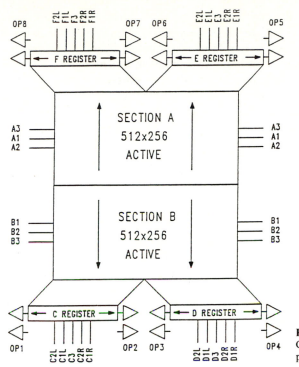

OP8   OP7 OP6   OP5

A3
A1
A2

SECTION A

512x256

ACTIVE

A3
A1
A2

B1
B2
B3

SECTION B

512x256

ACTIVE

B1
B2
B3

C REGISTER   D REGISTER

OP1   OP2 OP3   OP4

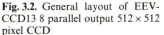

Fig. 3.2. General layout of EEV-CCD13 8 parallel output 512×512 pixel CCD

smaller number of pixels, for example, it is possible to image an area of 16–16 pixels at a sustained rate of 10 kHz with 12-bit (4096 grey level) digitisation, and a 64×64 pixel area at over 1 kHz (Fig. 3.1).

CCD manufacturers are also developing specialist CCDs optimised for a rapid read-out. Thomson (France) have a 14×14 pixel array that can be read out at over 20 kHz repetition rate. EEV (UK) have recently announced a 512 $\times 512$ pixel CCD that has 8 independent outputs (Fig. 3.2) and can be run up to around 30 MHz at each output. Such a device is capable of producing a sustained data output rate of as much as 0.25 gigabyte per second, enough to bring the most powerful computer system to its knees.

When the highest dynamic range is required much lower pixel rates are essential to allow high accuracy digitisation (16–20 bit, 65,000–1,000,000 grey levels). Although these are not usually considered as fast imaging systems they may also be operated in reduced format mode to give small format imaging repetition rates of tens or hundreds of images per second with the highest dynamic range.

5 Conclusions

The researcher looking for a fast high quality imaging system for fluorescence work has a variety of choices now open. The most flexible and probably the most cost-effective are those that use CCDs in a programmable, variable format manner. The current state of what can be achieved routinely is shown in Fig. 1 along with some of the latest devices carrying on the market. Rapid progress has been achieved in the performance of imaging systems in recent years and this will doubtless continue into the future.

4. Kinetic Studies on Fluorescence Probes Using Synchrotron Radiation

Wolfgang Rettig

I.N. Stranski Institute, Technische Universität Berlin, Strasse des 17. Juni 112, D-1000 Berlin 12, Germany

1 Introduction: Adiabatic Photochemical Product Formation and Fluorescence Probes

Many flexible aromatic molecules undergo spontaneous intramolecular rotational relaxation processes in the excited state leading to an energy minimum far away from the initial geometry which thus can be termed a photochemical product. Because the process occurs entirely in the excited state, it is called an "adiabatic photoreaction" [1]. An experimentally very well-studied example is the double-bond twisting of excited stilbene [2]. In recent years, the family of the Twisted Intramolecular Charge Transfer (TICT) compounds has been developed to a great extent [3–6]. In these compounds, two aromatic moieties are linked by a single bond, and excited-state rotational relaxation occurs towards a twisted conformation, coupled with intramolecular electron transfer (Fig. 4.1). Modern theoretical concepts can describe the twisting of both double and single bonds in one and the same model, that of biradicaloid states [6–8].

In favourable cases, the product of an adiabatic photoreaction is emissive, and thus its properties can be studied conveniently by time-resolved or time-integrated fluorescence spectroscopy. From the kinetic measurements, information can be drawn as to the medium influence on the reaction rate. Thus these dyes with a large-amplitude motion leading to an emissive photochemical product are well suited to act as fluorescence probe of the microenvironment, reporting both on its microviscous (kinetic measurements) as well as its micropolar properties (shift of the fluorescence spectrum). Even in cases where the TICT product is nonemissive, the primary excited state is quenched by the photoreaction, and its kinetics can still be followed by observing the precursor fluorescence.

Adiabatic photoreactions leading to a fluorescent product have the potential of transforming high-energy photons into low-energy ones, because the product emission is usually strongly redshifted with respect to the emission of the precursor. Therefore, absorption and emission hardly overlap, and reabsorption processes are minimized. This decoupling of absorption and emission is highly desirable for designing fluorescence probes used as labels, for example in fluoroimmunoassays [9], because it can remove the interference with scattered light and with fluorescent impurities which usually do not exhibit this strong redshift of the fluorescence. The TICT mechanism is not the only one

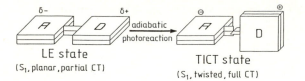

LE state
(S$_1$, planar, partial CT)

TICT state
(S$_1$, twisted, full CT)

Fig. 4.1. Grabowski's model of TICT formation [3]: An adiabatic photoreaction leading from an excited precursor with near-planar conformation and strong but incomplete charge transfer (locally excited "LE" or "B" state) to an excited product with twisted conformation and virtually complete charge separation ("TICT" state or "A")

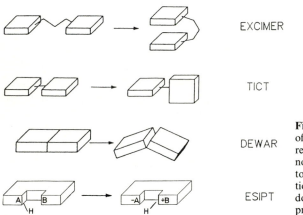

EXCIMER

TICT

DEWAR

ESIPT

Fig. 4.2. Schematic representation of four classes of adiabatic photoreactions with luminescent and nonluminescent products. From top to bottom, the necessary reaction volume decreases, and so does the sensitivity to the viscous properties of the medium

leading to strongly redshifted fluorescence from an adiabatic photoproduct. Other well-known examples are excimer or exciplex formation [10, 11] and intramolecular excited state proton transfer (ESIPT) [12] (Fig. 4.2).

These latter two types of photoreaction are extremes regarding the sensitivity to the surrounding: Excimer formation, involving translational components in the movement of relatively bulky aromatic moieties, is very easy to suppress by high-viscosity conditions of the surrounding [13], whereas ESIPT involves mainly the movement of a proton and thus cannot be stopped even in the most viscous environments like low-temperature frozen solvents or rigid polymers [14]. Intermediate in this respect is TICT formation which can be used to probe microviscosity of liquids around the glass transition temperature [13]. A special but very useful adiabatic photoreaction is the formation of Dewar anthracenes from anthracenes with very bulky substituents [15]. In this case, the adiabatically reached product (a photochemical funnel in the S$_1$ surface) is nonemissive, but fluorescence quenching of the precursor occurs and can be used to probe the free volume of rigid polymer matrices well below the glass transition temperature, because the reaction requires even less space than TICT formation. Figure 4.2 summarizes the most common adiabatic photoreactions, ordered according to the necessary reaction volume.

Dyes which exhibit these photoreactions, lend themselves to various applicational aspects: As already outlined, they can be advantageously used for probing the microviscous properties of the surrounding medium, both in polymer and material science [13] and in biology. The sensitivity of the TICT compounds can simultaneously be exploited in their use as polarity probes. It will be shown below that in actual fact, most of the fluorescence probes used in biology [16] are of this type. The large Stokes shift of these compounds can, on the other hand, be used for fluorescence labelling purposes, but it can also be used to develop more efficient fluorescent solar collectors [17]. TICT reactions are especially easy to modify by substituent effects. By choosing substituents with several functions (TICT donor but at the same time ion complexing or basic center), it is also possible to develop analytical ion or pH sensing fluorescence probes.

The remainder of this paper will focus mainly on TICT states. Its principles will be outlined in more detail, and various ways will be shown how TICT fluorescence can be manipulated by molecular design and how microviscosity, micropolarity and analytical probes can be developed. A final chapter will deal with the properties of proton transfer dyes.

2 Fluorescence Kinetics and Its Application to TICT States

2.1 The Principle of TICT States

Because they involve virtually pure charge transfer, the energy of the TICT states is governed by the electron donor and electron acceptor properties of the subsystems. A convenient measure of these quantities are the ionization potential (IP) and the electron affinity (EA), or oxidation and reduction potentials, and Eq. 1 can be used to quantify the TICT energy.

$$E(\text{TICT}) = \text{IP(donor)} - \text{EA(acceptor)} + C + \Delta E_{\text{solv}}. \tag{1}$$

C and ΔE_{solv} are correction factors regarding the Coulombic stabilization by radical anion/radical cation attraction in the TICT state, and describing the solvent stabilization energy [4].

Although for every twisted bichromophoric system, several TICT states exist corresponding to transitions from different donor to different acceptor orbitals, TICT formation occurs spontaneously only if it is exothermic in S_1, i.e. if the energy of the lowest TICT state is below that of the planar LE state, i.e. if inequality Eq. 2 holds.

$$E(\text{TICT}) < E(\text{LE}). \tag{2}$$

Usually, $E(\text{LE})$ depends much more weakly on IP and EA than $E(\text{TICT})$ such that by changing IP and/or EA of one or the other subsystem, the driving force for TICT formation can be tuned, and TICT formation can be switched on and off.

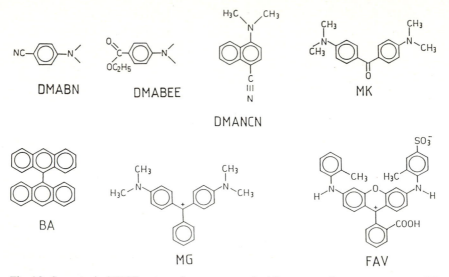

Fig. 4.3. Some typical TICT systems. In many cases, dual fluorescence from precursor state LE and product TICT is observed simultaneously

Figure 4.3 shows some representative examples of TICT molecules. A larger list of compound families can be found in Refs. [4, 8].

The most important properties of TICT states can be summarized as follows:
– Normally, the energy minimum of the TICT state occurs for the perpendicular conformation, such that the π-systems of donor and acceptor are perpendicular and decoupled from each other.
– Because of this decoupling, the fluorescence from TICT states is "forbidden", i.e. weak unless other effects like vibronic coupling with allowed states counteract (which is often the case).
– Triplet and Singlet TICT states are expected to be nearly degenerate.
– TICT formation kinetics can be governed by tuning the TICT energy (Eqs. 1, 2) or by molecular modelling, e.g. chemical bridging of the twisting moieties (rigidization) or pretwisting by sterical hindrance (e.g. *ortho*-alkyl substitution), or by incorporation in a rigid matrix (low-temperature glass or polymer).

The reason for an energy minimum in S_1 and a maximum in S_0 for the perpendicular TICT energy is the interaction of the ground with the excited state which occurs more strongly for deviations from perpendicularity [18].

2.2 The Time Structure of Synchrotron Radiation

The Berlin electron synchrotron BESSY uses mainly two modes of operation, the multi-bunch mode consisting of a series of pulses spaced by 2 nsec, with about 300 ps halfwidth, and the single-bunch mode where the pulses are spaced

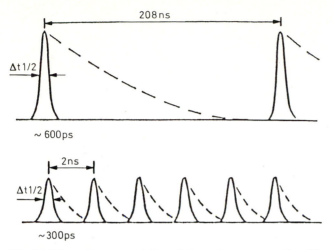

Fig. 4.4. Schematic representation of the pulse sequence of multi- and single-bunch modes at BESSY

by 208 ns (halfwidth around 500–600 ps [19, 20] and are extremely reproducible in shape (Fig. 4.4).

This time structure is available simultaneously in a very broad wavelength range extending from Far Infrared over visible/UV to vacuum UV and X-ray and can be used to achieve a considerable reduction of noise or background in the experiments [20]. It can also be used to follow fast kinetic events directly, and one of the possibilities is to use a time-correlated single photon counting equipment to measure fluorescence decays with a time resolution down to 100 ps [21, 22]. An example is given in Fig. 4.5 which compares the dual fluorescence spectra and decay traces of the typical TICT compounds DMABN and DMABEE, the nitrile and its corresponding ethyl ester derivative. As can be seen, the ester exhibits an increased fraction of TICT fluorescence for comparable conditions (Fig. 4.5a), which is paralleled by a faster decay of the precursor state LE (short wavelength fluorescence) (Fig. 4.5b) and a corresponding faster rise time of the TICT fluorescence [23–25]. Thus, the TICT formation kinetics in the ester is much faster than in the nitrile although the rotating moieties do not differ strongly in rotational volume. The reason is that the driving force or, more exactly speaking, the topology of the reactive hypersurfaces differ, possessing a "conical intersection" for the nitrile (a kinetic hindrance) which is absent for the ester (Fig. 4.5c) [10, 25, 26].

3 How to Construct TICT Fluorescence Probes

3.1 Principal Approaches

The dual fluorescence as shown in Fig. 4.5 can be developed into two directions: By choosing more and more favourable energetic conditions for TICT formation, the excited state equilibrium is shifted towards the product state, such that

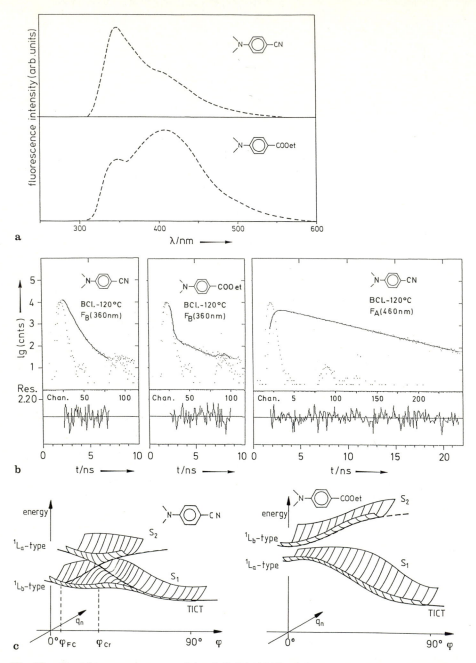

Fig. 4.5. a Dual fluorescence spectra of the nitrile DMABN and the corresponding ester DMABEE in *n*-butyl chloride at room temperature; **b** fluorescence decay traces at $-120°C$ of the short wavelength bands (F_B, LE state) of these compounds, and rise and decay curve for the TICT fluorescence (F_A) of DMABN. The rise time of DMABN matches the decay time of the short wavelength fluorescence (440 ps), whereas the corresponding kinetics are much faster in the ester (< 100 ps) [23, 26]; **c** artist's view of the S_1 and S_2 hypersurfaces for DMABN (with conical intersection) and DMABEE (without) [4, 25]

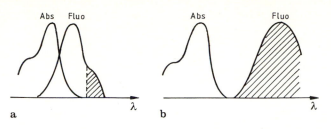

Fig. 4.6. Schematic absorption and fluorescence spectrum **a** of a dye with normal Stokes shift, **b** of a dye where an adiabatic photoreaction produces an anomalously large Stokes shift. In the first case, under high-concentration or long-pathlength conditions, most of the emitted fluorescence (*unshaded area*) is reabsorbed and partly lost

in the limit only the TICT fluorescence is observed. This leads to a large separation of absorption and emission bands (Stokes shift) and is an important property if reabsorption processes have to be avoided (Fig. 4.6), such as in the development of more efficient fluorescent solar collectors [17]. A large Stokes shift is also helpful in fluorescence labelling techniques, because complications by fluorescent impurities or by scattered light can thus be circumvented [9]. In this way, the signal to background ratio and therefore the sensitivity can be significantly enhanced. A drawback in using TICT fluorescence compounds is their intrinsically weak fluorescence (see above), which is, however, often counteracted by vibronic coupling. Thus, many of the examples shown below show TICT fluorescence quantum yields of 0.4 and more. Another approach is to use different types of adiabatic photoreactions, such as ESIPT, where product emission is not intrinsically forbidden (see last section).

The second possibility in constructing TICT probes is to use nonemitting TICT states as intramolecular fluorescence quenching channels. As the TICT energy is controllable by donor–acceptor energetics, the driving force for TICT formation is similarly controlled, and therefore the quenching channel of the short wavelength fluorescence can be switched on and off, either as a function of molecular structure (design of highly fluorescing or of fast relaxing fluorescence dyes with small Stokes shift) or in response to changes in the surrounding (fluorescence probes).

3.1.1 Fluorescing Dyes with a Large Stokes Shift

Figure 4.7 exemplifies how introduction of acceptor substituents can entail TICT fluorescence properties. In highly polar solvents like acetonitrile, the structured short-wavelength band has completely disappeared, and only the strongly redshifted TICT band is present, with a quantum yield of about 0.4 [27].

Figure 4.8 collects the structures of other efficiently fluorescing TICT compounds. Related to CBN is the indole IBN and the pyrrole PBN. Although the latter is less efficient in fluorescence, it has been shown to be well suited for

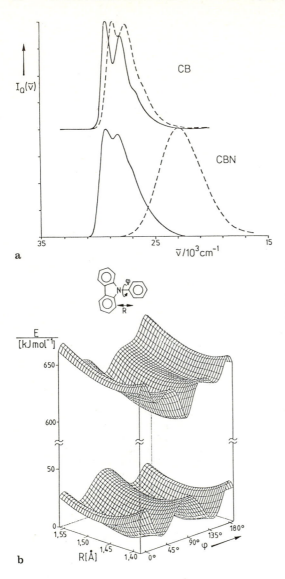

a

b

Fig. 4.7. a Fluorescence spectra of *N*-phenylcarbazole (CB) and its *p*-cyano derivative (CBN) in *n*-hexane (———) and acetonitrile (– –) [27]. CB shows normal B-type fluorescence in both solvents, whereas CBN shows pure TICT fluorescence in the polar environment where polar states are preferentially stabilized. **b** Quantum chemical force field calculation for the ground and the TICT state of CB [27]. The TICT state possesses a minimum for the 90° twisted conformation where in the ground state the top of the rotational barrier is located. In CB, the TICT state is a higher excited state, and is not polulated photochemically. In CBN, it is lowered energetically relative to the locally excited states (which show topologies similar to the ground state), and the minimum at 90° can be polulated. Note that the ground state equilibrium twist angle is around 40° which corresponds to pretwisting and enhances the TICT formation rate

engineering the size of the Stokes shift by varying the donor ionization potential. Thus, a PBN derivative has been studied where absorption occurs below 300 nm, and emission is redshifted to about 420 nm in nonpolar hexane and to 630 nm in polar acetonitrile [28]. In mixed solvents of intermediate polarity, PBN shows dual fluorescence (Fig. 4.9). This can be used in principle to detect traces of water or other polar impurities in nonpolar solvents. The sensitivity is rather high because preformed ground-state clusters are involved (static quenching).

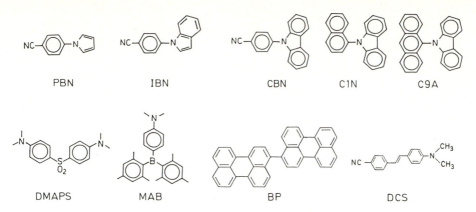

PBN   IBN   CBN   C1N   C9A

DMAPS   MAB   BP   DCS

Fig. 4.8. Structure of TICT compounds with largely Stokes-shifted fluorescence bands

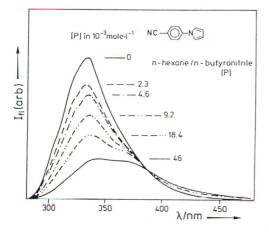

Fig. 4.9. Quenching effect of the short wavelength fluorescence and corresponding built-up of the TICT fluorescence of PBN by adding traces of a polar solvent component (here *n*-butyronitrile) to a hexane solution [5]. Note that not only the intensity but also the band shape changes. The latter is less sensitive to complications arising from scattering and impurities

Well-known laser dyes like DCM probably also fluoresce from a TICT state. This can be concluded from a comparison with the closely related dimethyl-amino-cyano-stilbene DCS (and its bridged derivatives [18, 29]) and is related to the strong solvent polarity sensitivity of the fluorescence yield of such dyes. Some symmetric biaryls also show dual fluorescence connected with TICT formation. One very recent example of a highly emissive TICT biaryl absorbing above 400 nm is biperylenyl (BP) [30].

3.1.2 Probes with a Highly Sensitive Intramolecular Fluorescence Quenching Process

Many well-known dyes like rhodamines are ionic. TICT states in ionic dyes are usually nonemissive because they are linked not with charge separation but with charge shift. TICT formation in these dyes therefore acts as nonradiative decay

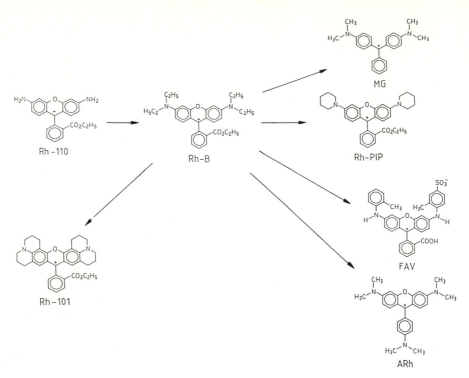

Fig. 4.10. Various possibilities of introducing a TICT quenching channel into dyes related to Rhodamine 110. The importance of the nonradiative channel is indicated by the compound's position towards the right

process and quenches the normal fluorescence. As a well-known example, different TICT formation possibilities are shown in Fig. 4.10 starting from a TICT inactive Rhodamine dye (the laser dye Rhodamine 110 bearing amino substituents).

Taking away the oxygen bridge (Malachite Green MG) opens up the possibility of twisting one or two anilino groups leading to TICT-like states with charge localization: Very strong fluorescence quenching occurs [22–24, 31].

But enhancing the donor capacity of the amino groups themselves, by dialkyl substitution, also generates nonradiative TICT channels linked with rotation around a different bond. These channels are, however, much less efficient than in the case of triphenylmethane dyes. Thus, the laser dye Rhodamine B (diethylamino substituents) shows sizeable fluorescence quenching especially in highly polar environments and at high temperatures [33]. Julolidine-type bridging as in Rhodamine 101 suppresses this channel, off course. Pretwisting (Rh-PIP) enhances it [33]. These general rules also work for other related dye families like oxazines and thiazines [34, 35]. If the donor strength of the dialkylamino groups is further enhanced, by aryl substitution as in Fast Acid Violet (FAV), Rhodamine dyes are derived with very short fluorescence lifetimes due to efficient intramolecular rotation and TICT formation [36]. They can be used for probing the microviscous solvent properties.

Finally, the flexible carboxy-phenyl group in Rhodamines, behaving as electron acceptor, can also be modified to act as donor by introducing an amino or dialkylamino group. Thus, the compound ARh in Fig. 4.10 shows a very efficient quenching channel due to TICT formation. This property proves to be a handle which allows to develop analytical fluorescence probes.

Two examples are shown in Fig. 4.11. On protonating ARh, the dialkylamino group on the flexible phenyl ring is transformed into an ammonium group which acts as a weak electron acceptor, similar to the carboxy group in ordinary Rhodamines. The corresponding TICT energy is raised, and the fast quenching channel (k_{T1}) is closed. Therefore, protonation of ARh leads to a strong increase of the fluorescence quantum yield (by a factor of more than 100 [37]). Only the much slower TICT channel k_{T2} also present in Rhodamine B etc. remains. ARh lends itself for the possibility of very sensitive fluorescence titration and for pH determination in very small volumes like the interior of cells etc. Figure 4.11 also shows a corresponding acridinium compound with similar properties in response to acids, which has been further developed into an ion sensing fluorescence probe: The crown ether function complexes ions like Ag^+ and partly

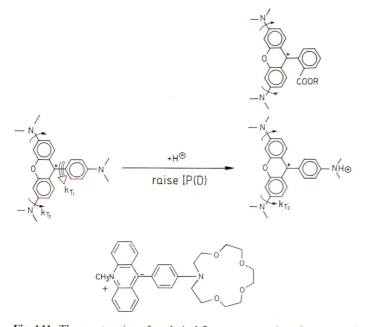

Fig. 4.11. The construction of analytical fluorescence probes: On protonation, the flexible anilino group with good donor properties in ARh is transformed into an anilinium moiety unable to act as TICT donor, similar to the phenylcarboxyester group of ordinary Rhodamine dyes. ARh/H^+ shows high fluorescence quantum yields and long lifetimes, whereas the fluorescence of ARh is strongly quenched. A similar effect occurs in the acridinium derivative shown. Here, the crown ether function allows not only to detect protons, but also metal ions

uses the lone pair of the amino group. This reduces its donor properties and closes the TICT quenching channel [38, 39].

4 TICT Probes in Biology

Compounds like anilino-naphthalene sulfonates (ANS, DNS, TNS) and PRODAN (Fig. 4.12) are popular fluorescence probes for sensing the polarity of the microenvironment in biological materials [16]. Their sensitivity has been recognized to be due to the formation of fluorescent TICT states [40, 41]. In favourable cases, even clearly separated dual fluorescence is observable (Fig. 4.13). This could in principle be exploited to develop more versatile fluorescence probes because not only the wavelength but also the relative intensity is available and changes as a function of the parameters of the

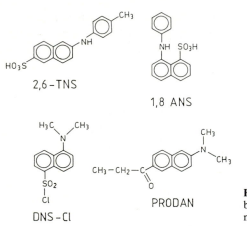

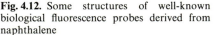

Fig. 4.12. Some structures of well-known biological fluorescence probes derived from naphthalene

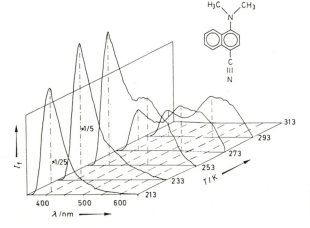

Fig. 4.13. Dual fluorescence of DMANCN in propylene glycol as a function of temperature [56]

surrounding. Quantum chemical methods can likewise be used to predict structures of new biological fluorescence probes [42].

The relaxation of TICT molecules can furthermore serve to probe more complex properties of environments. For example, 9,9'-bianthryl proves to be a probe which allows to measure the solvation rate i.e. the effective reorientation time of the individual solvent dipoles [43–45]. For many but especially for highly viscous solvents, a single time constant is not sufficient, and distributions of reorientation relaxation times are better used to interpret the data. Pressure dependent measurements can yield information on how these distributions change [45], and a similar approach can in principle be used to probe the relaxation behaviour of biological material.

5 Application of TICT and Other Relaxing Probes in Liquid and Rigid Media

TICT formation is normally linked with a molecular twisting rearrangement (large-amplitude motion) (Fig. 4.1) which needs a certain reaction volume. In low-viscosity solvents at room temperature, fast solvent fluctuations provide this necessary volume. In rigid environments, however, only a very minor fraction of the molecules possess a neighborhood allowing for large-amplitude motion, and therefore, TICT formation is effectively stopped in rigid low-temperature glasses. A certain fraction of "free volume" is, however, always present and is responsible, for example, for the thermal expansion, the diffusive and the viscous properties of the medium. If this fraction is sizeable, it can be used by the TICT molecule. This is the reason why some TICT fluorescence can be observed even in rigid polymers at room temperature [46].

The Williams–Landel–Ferry (WLF) theory allows to quantify the free volume above the glass transition temperature using available steady-state or time-resolved fluorescence data. The free volume sensed depends on the reaction volume which the probe needs. More bulky excimer fluorescence probes (Fig. 4.2) intrinsically sense a smaller free volume than TICT probes [13]. But each probe can be used to compare different media. Thus, the free volume fractions of polydimethylsiloxane oils have been compared to that of ordinary alkane solvents, and it has been found (Table 4.1) that alkanes possess a considerably smaller free volume than polymers and that a factor of 10 difference in the macroscopic polymer viscosity is not linked to a change of the size of the free volume fraction but more likely to a different polymer entanglement.

Adiabatic photoreactions which require only a very small reaction volume can even be used to monitor effects below the glass transition temperature. An example is 10-cyano-9-tert-butyl-anthracene (TB9ACN) which reacts photo-chemically towards a Dewar isomer (Fig. 4.2). This process is linked with fluorescence quenching and occurs even in rigid polymethylmethacrylate (PMMA) at room temperature and below. The resulting shortened fluorescence lifetimes can be monitored using synchrotron radiation, and the decays turn out to be highly

Table 4.1. Free volume fractions $f_g = v_{free}/v_{total}$ at the glass transition temperature T_g derived from the intercept of WLF plots, for the TICT probe PYREST and the excimer probe DIPHANT in polydimethylsiloxane oils of different macroscopic viscosity and in acyclic and cyclic alkane solvents [13]

	PDMS S100[c]	PDMS S1000[d]	3MP[a]	MCH/MCP[b]
TICT probe	0.08	0.08	0.016	0.06
Excimer probe	0.05	0.05	—	—

[a] 3-methylpentane
[b] methylcyclohexane/methylcyclopentane
[c] macroscopic viscosity of 100 mPa s at room temperature
[d] 1000 mPa s

Table 4.2. Gaussian analysis of the decay time distribution of TB9ACN in PMMA at different temperature [13]

Temp/K	Most probable lifetime (ns)	Distribution halfwidth (ns)
261	7.3	7.3
243	8.7	7.2
224	10.5	7.2
203	11.7	6.8
175	13.6	3.3
146	15	1

nonexponential indicating the emission from an ensemble of molecules with different decay times. This decay time distribution is related to the distribution of free volume around the molecule. Table 4.2 summarizes the results [13]. The broadened lifetime distributions at higher temperature reflect a broadened free volume distribution.

Several other TICT molecules have been used to sense free volume properties of polymers and molecular liquids. Thus triphenylmethane dyes prove to be usable both above and below the glass transition temperature of the solvent [22, 47].

In a different approach, TICT molecules can be chemically linked to the polymeric backbone by spacers of different length thus allowing a detailed description of the mobility of polymer segments (see for example [48]; see also [49] for a review dealing with TICT molecules as fluorescent probes in polymers and other environments).

On surfaces, too, dual fluorescing compounds can be used as "reporters" for the surface structure and mobility. DMABN has been used on glass [50] and on metal oxide powders [51], and from the dual fluorescence observed even at 77 K (Fig. 4.14), it could be concluded that intramolecular rotational motion occurs relatively unhindered on the surface.

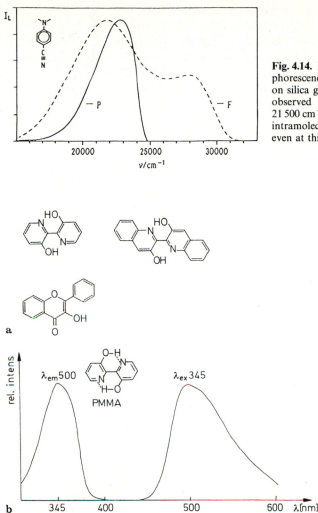

Fig. 4.14. Fluorescence (*F*) and phosphorescence (*P*) of **DMABN** adsorbed on silica gel powder, at 77 K [51]. The observed TICT fluorescence band at $21\,500\ cm^{-1}$ indicates high freedom of intramolecular rotational movement even at this low temperature

Fig. 4.15. a Chemical structure of highly fluorescent ESIPT compounds; **b** completely decoupled absorption and emission of $BP(OH)_2$ in a PMMA matrix at room temperature

6 Proton Transfer Fluorescence Probes

The excited state intramolecular proton transfer (ESIPT) (Fig. 4.2) is an adiabatic photoreaction especially efficient in producing large Stokes shifts and separating absorption from emission regions. Moreover, as it necessitates only a very small reaction volume, it occurs efficiently also in surroundings which prevent large-amplitude motion. Fig. 4.15 shows as an example absorption and fluorescence spectrum of an ESIPT active biphenyl derivative. It is highly emissive [14] and highly photostable [52] and can even be used as a laser dye [53], similar to the well known ESIPT-dye 3-hydroxy-flavone [54]. It could be used for labelling applications because of its large Stokes shift and corresponding insensitivity to fluorescing impurities, and for constructing more efficient fluorescent solar collectors with diminished reabsorption (see Fig. 4.6) [17]. This class of compounds is open for various developments. For example, coumarine- or chromone-based ESIPT dyes have been predicted by quantum chemical calculations with absorption spectra in the visible and even the near-infrared range [55].

Acknowledgement. The results outlined above have become possible due to support by BMFT (fluorescence kinetics using BESSY synchrotron radiation, projects 05 314 FAI5 and 05 414 FAB1), the Techn. Univ. Berlin (Interdisciplinary Project IFP4) and the DFG (Heisenberg fellowship to the author), and due to the skill of my coworkers, in particular Dr. M. Vogel, Dr. W. Majenz, R. Fritz, A. M. Klock, D. Braun, F. Vollmer and Mrs. B. Paeplow.

7 References

1. Turro NJ, McVey J, Ramamurthy V, Lechtken P (1979) Angew Chem 91: 597
2. Waldeck DH (1991) Chem Rev 91: 415
3. Grabowski ZR, Rotkiewicz K, Siemiarczuk A, Cowley DJ, Baumann W (1979) Nouv J Chim 3: 443
4. Rettig W (1986) Angew Chemie 98: 969; Angew Chem Internat Ed Engl 25: 971
5. Lippert E, Rettig W, Bonačić-Koutecký V, Heisel F, Miehé JA (1987) Adv Chem Phys 68: 1
6. Rettig W (1991) EPA newsletter 41: 3
7. Michl J, Bonačić-Koutecký V (1990) Electronic aspects of organic photochemistry, Wiley, New York
8. Rettig W (1988) In: Liebman J, Greenberg A (eds) Modern Models of Bonding and Delocalization. VCH-Publishers, New York, p 229 (Molecular Structure and Energetics vol 6)
9. Soini E, Hemmilä I (1979) Clin Chem 25: 353
10. Birks JB (1970) Photophysics of aromatic molecules. Wiley, London, chap 7
11. Beens H, Weller A (1975) In: Birks JB (ed) Organic molecular photophysics, vol 2. Wiley, London, chap 4
12. Grabowska A, Waluk J, Bulska H, Mordzinski A (1986) Nouv J Chim 10: 413
13. Rettig W, Fritz R, Springer J (1991) In: K. Honda, N. Kitamura, H. Masuhara, T. Ikeda, M. Sisido, MA Winnik (eds) Photochemical Processes in Organized Molecular Systems, Elsevier, Amsterdam
14. Bulska H (1988) J Lumin 39: 293
15. Jahn B, Dreeskamp H (1984) Ber Bunsenges Phys Chem 88: 42
16. Lakowicz JR (1983) Principles of Fluorescence Spectra. Plenum, New York
17. Rettig W (1991) Nachrichten aus Chemie Technik Laboratorium 39: 398
18. Rettig W, Majenz W, Lapouyade R, Haucke G (1992) J Photochem Photobiol A: Chemistry 62: 415
19. Rettig W, Vogel M, Klock A (1986) EPA newsletter 27: 41
20. Rettig W, Wiggenhauser H, Hebert T, Ding A (1989) Nuclear Instruments and Methods in Physics Research, A277: 677
21. O'Connor DV, Phillips D (1984) Time-correlated Single Photon Counting. Academic, London
22. Vogel M, Rettig W (1987) Ber Bunsenges Phys Chem 91: 1241
23. Rettig W, Vogel M, Lippert E, Otto H (1986) Chem Phys 103: 381
24. Wermuth G, Rettig W, Lippert E (1981) Ber Bunsenges Phys Chem 85: 64
25. Rettig W, Wermuth G (1985) J Photochem 28: 351
26. Rettig W (1992) In: Malaga N, Okada T, Mosukara H (eds) Dynamics and Mechanisms of Photoinduced Electron Transfer and related Phenomena. Elsevier, Amsterdam, p. 57
27. Rettig W, Zander M (1982) Chem Phys Lett 87: 229

28. Rettig W, Marschner F (1990) New J Chem 14: 819
29. Rettig W, Majenz W (1989) Chem Phys Lett 154: 335
30. Rettig W, Majenz W, Lapouyade R, Vogel M (1992) Proceedings Ibaraki Symposium, Tsukuba/Japan, May 1991, J Photochem Photobiol. A: Chemistry, in press
31. Vogel M, Rettig W (1985) Ber Bunsenges Phys Chem 89: 962
32. Rettig W (1988) Appl Phys B45: 145
33. Vogel M, Rettig W, Sens R, Drexhage KH (1988) Chem Phys Lett 147: 452
34. Vogel M, Rettig W, Sens R, Drexhage KH (1988) Chem Phys Lett 147: 461
35. Vogel M, Rettig W, Fiedeldei U, Baumgärtel H (1988) Chem Phys Lett 148: 347
36. Tredwell CJ, Osborne AD (1980) J Chem Soc Faraday II, 76: 1627; Osborne AD, Winkworth AC (1982) Chem Phys Lett 85: 513
37. Vogel M (1987) PhD thesis, Berlin
38. Jonker SA, Ariese F, Verhoeven JW (1989) Recl Trav Chim Pays-Bas 108: 109
39. Jonker SA, Verhoeven JW, Reiss CA, Goubitz K, Heijdenrijk D (1990) Recl Trav Chim Pays-Bas 109: 154
40. Cowley DJ (1986) Nature 319: 14
41. Kosower EM, Huppert D (1986) Ann Rev Phys Chem 37: 127
42. Nowak W, Rettig W, J Mol Struct THEOCHEM, submitted
43. Kang TJ, Jarzeba W, Barbara PF, Fonseca T (1990) Chem Phys 149: 81
44. Mataga N, Yao H, Okada T, Rettig W (1989) J Phys Chem 93: 3383
45. Lueck H, Windsor MW, Rettig W (1990) J Phys Chem 94: 4550
46. Al-Hassan KA, Rettig W (1986) Chem Phys Lett 126: 273
47. Anwand D, Müller FW, Strehmel B, Schiller K (1991) Makromol Chem 192: 1981
48. Tazuke S, Guo RK, Ikeda T, Ikeda T (1990) Macromolecules 23: 1208
49. Rettig W, Baumann W (1992) In: Rabek JF (ed) Photochemistry and Photophysics, Vol. VI. CRC Press, in press
50. Levy A, Avnir D, Ottolenghi M (1985) Chem Phys Lett 121: 233
51. Günther R, Oelkrug D, Rettig W (1991) to be published
52. Grabowska A, Bulska H, Rymarz G (1987) Polish Patent No 266542
53. Sepioł J, Bulska H, Grabowska A (1987) Chem Phys Lett 140: 607
54. Kasha M, Aartsma TJ, McMorrow D, Chou T (1984) J Phys Chem 88: 4596
55. Vollmer F, Rettig W (1990) unpublished results
56. Ayuk AA, Rettig W, Lippert E (1981) Ber Bunsenges Phys Chem 85: 553

5. Dynamics and Geometry in Dimeric Flavoproteins from Fluorescence Relaxation Spectroscopy

Philippe I. H. Bastiaens, Antonie J. W. G. Visser

Department of Biochemistry, Agricultural University, Dreijenlaan 3, NL-6703 HA Wageningen, The Netherlands

1 Introduction

The biologically widespread group of flavoproteins have in common that they contain the yellow flavin molecule as prosthetic group. The most common natural flavins are flavin adenine dinucleotide (FAD) and flavin mononucleotide (FMN) which have both riboflavin as their biological precursor. To the biological chemist the most interesting feature of this versatile molecule is its redox properties which can be modulated by the (protein) environment. The redox active part of the natural flavins is the isoalloxazinic ring which can exist in the oxidized, one-electron reduced and two-electron reduced state. The redox potentials of the two one-electron steps vary greatly among different flavoproteins and depend on the chemical nature of the active site in which the isoalloxazine resides. This property makes flavin suitable as an electron shuttle in very different chemical (redox) reactions which explains its widespread occurrence in nature [1].

For the biophysicist one of the most interesting properties of flavin is that it is a fluorescent molecule whose spectral parameters depend on the particular characteristics of the molecular environment. In this respect the molecule is a natural reporter group from which information on the dynamical structure of proteins in solution can be extracted. The large variety of flavoproteins with their different biological functions provides a pool of material to undertake comparative studies on protein dynamics and structure in relation to catalytic function. Most of the fluorescence data are obtained by monitoring the fluorescence of the intrinsic probe tryptophan [2]. Although this amino acid is present in a diversity of proteins, several distinct disadvantages adhere to the use of this molecule as a natural reporter group. First, most proteins contain several tryptophan residues which makes the interpretation of the data ambiguous because of the different contributions to the fluorescence from differently emitting classes of tryptophan [3]. Second, the photophysics of tryptophan is complicated because of the existence of almost degenerate energy levels belonging to the first absorption band [4]. Due to this latter property the fundamental polarization of this molecule is considerably lowered by interconversion between the two perpendicular transition moments [5]. Also, the intramolecular excited-state process considerably complicates the interpretation

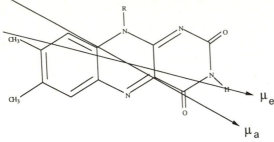

Fig. 5.1. Molecular structure of flavin with directions of absorption (μ_a) and emission (μ_e) transition moments

of the data in the picosecond time domain where rapid fluctuations of protein residues can take place [6, 7].

In comparison, the photophysical properties of the isoalloxazinic residue of natural flavins are somewhat simpler. The first electronic transition of free flavin in neutral aqueous solution has an absorption maximum at 445 nm ($\varepsilon = 12.4 \, \text{mM}^{-1} \, \text{cm}^{-1}$) and the second electronic transition has an absorption maximum at 367 nm ($\varepsilon = 10.0 \, \text{mM}^{-1} \, \text{cm}^{-1}$) [8]. The geometry of the first electronic transition moment in the molecular frame is known (see Fig. 5.1) [9]. From the fundamental anisotropy it was deduced that the emission transition moment makes an angle of 15° relative to the first absorption transition moment. Although two possibilities remain for the relative orientation of the emission transition moment, we determined that it lies almost parallel to the pseudosymmetry axis of the molecule [10].

The flavins fluoresce in the green spectral region where the position of the emission maximum is dependent on the dielectric and refractive properties of the solvent. The fluorescence quantum yield of riboflavin or FMN ($Q = 0.26$) in aqueous solution is much higher than that of free FAD ($Q = 0.03$) [11]. The quenching of the isoalloxazine fluorescence in FAD originates from intra-molecular complexation with the adenine moiety where both static and dynamic mechanisms account for the decrease in the quantum yield relative to FMN. Detailed thermodynamic and kinetic information has been obtained on this complexation from steady-state and time-resolved fluorescence studies [12]. The fluorescence of flavin bound to apo-flavoproteins is in most cases rather weak which in some instances can be ascribed to a transient charge transfer mechanism with electron-rich aromatic residues (tyrosine, tryptophan) [13]. Although exceptions such as lipoamide dehydrogenase and electron transferring flavoprotein with high quantum yield exist, the weak fluorescence of flavopro-teins demands sensitive detection systems to obtain accurate information on dynamics and structure of flavoproteins [14]. The advantage of the green fluorescence of flavins is that the signal is much less contaminated by fluor-escence of (protein) impurities. However, extreme care should be taken to ascertain that the sample does not contain free flavin which can arise from the dissociation of the prosthetic group from the apo-protein.

Modern fluorescence techniques provide a vast arsenal to obtain information on structure and dynamic aspects of proteins in solution. The information that the (biological) fluorescence spectroscopist can retrieve from experiments include: hydrodynamic properties of macromolecules (e.g. shape and size), thermodynamic parameters of protein–protein and ligand–protein equilibria, kinetics and thermodynamics of conformational substates, flexibility in proteins, dipolar relaxation and estimation of polarity inside proteins, distances between chromophoric groups (geometric information) and kinetic parameters of excited-state reactions.

2 Lipoamide Dehydrogenase and Glutathione Reductase as Biophysical Systems

The structurally similar dimeric flavoproteins glutathione reductase (GR) and lipoamide dehydrogenase (LipdH) contain one FAD per subunit of 50 kDa molecular mass. Both enzymes have been characterized in considerable detail [15–21]. The chemical mechanism of the enzymatic catalysis has been elucidated to a large extent. The enzymes have a redox-active disulfide group in common. LipdH functions within a multienzyme complex (e.g. the pyruvate dehydrogenase complex) [22]. The enzyme catalyses the oxidation of dihydrolipoamide coupled to the reduction of NAD^+:

$$Lip(SH_2) + NAD^+ \rightarrow LipS_2 + NADH + H^+$$

GR operates as a dimer in erythrocytes where it maintains a high value for the ratio of reduced to oxidized glutathione concentrations which is, among others, essential for proper function of erythrocytes by stabilizing thiols in the cell membrane. The enzyme utilizes NADPH instead of NADH and, in contrast to LipdH, oxidizes the cofactor [19]:

$$GSSG + NADPH + H^+ \rightarrow 2GSH + NADP^+$$

The crystal structures of GR from human erythrocytes and LipdH from *Azotobacter vinelandii* have been solved to 1.54 Å and 2.2 Å, respectively [23–25]. Notwithstanding the 28% homology the three-dimensional structure is very similar. The gene of lipoamide dehydrogenase from *Azotobacter vinelandii* (LipdH-AV) has been cloned and brought to expression in *E. coli* TG2 [26]. In this way the enzyme can be overproduced. More important, the active site can be modified by site-directed mutagenesis [27]. This provides not only a way to study the role of amino acid residues in catalysis, but also to modulate physical factors (electrostatics and statistical mechanics) which have an influence on conformational dynamics.

In studying the fluorescence properties of GR and LipdH two viewpoints can be distinguished. From a biological point of view it is striking that two proteins with very similar tertiary structure not only catalyze different reactions, but also operate in different supramolecular organizations. It is then of interest to use spectroscopic methods to investigate the role of conformational dynamics in relation to enzymatic function. From a physical viewpoint, however, the unique spatial arrangement of the flavins in conjunction with the specific local environment, permits the development and testing of (photo)physical models. With site-directed mutagenesis we are then able to alter the local chromophore environment and to trace its effect on the developed model. Frauenfelder has considered a protein as a physics laboratory [28], which is very appropriate in this context. One of the physical phenomena that can be investigated in these oxidoreductases is radiationless energy transfer between like chromophores (flavins), in relation with excitation energy and dipolar relaxation [29–31].

LipdH is one of the most fluorescent flavoproteins known with a flavin fluorescence quantum yield of about 0.1. In contrast, GR has very weak fluorescence ($q \approx 0.01$) [32]. A tyrosine (Tyr197) in close proximity of the isoalloxazine of FAD is responsible for a charge transfer exciplex reducing the fluorescence lifetime to the picosecond regime [33]. This tyrosine is absent in LipdH explaining its relatively strong fluorescence. The average fluorescence lifetimes of the two proteins follow the same tendency as their quantum yields. The polarized fluorescence decays are also widely different in both proteins. The decays as analyzed by a discrete set of exponentials exhibited a complex kinetic pattern [32]. With sophisticated apparatus and modern methods of data analysis we are now able to explain the experimental results in an unambiguous way.

In the following section we will focus attention on three methodologies in fluorescence spectroscopy. From each method specific information on the protein system can be obtained.

3 Fluorescence Methodology

3.1 Molecular Relaxation Spectroscopy

The interaction of a chromophore with the (dipolar) environment is character-ized by the action of two forces: (i) the polarizing force leading to the generation of the reactive field (R) and (ii) the restoring force tending to restore the configuration of the environment without the fluorescent molecule. The ex-pression relating the free energy of the solvate (chromophore surrounded by solvent or protein dipoles) to the reactive field has a parabolic form for both ground- and excited states [34]. This is schematically depicted in Fig. 2. Excitation in the main-band of the absorption spectrum with photons of frequency ν_m results in the population of non-equilibrium excited states. Excita-tion at the red edge of the absorption band (frequency ν_r) photoselects chromophores with the smallest electronic transition frequencies [30, 35]. Both excited-state solvates, obtained after red-edge and main-band excitation, will have a tendency to relax to the minimal energy configuration with a character-istic dipolar relaxation time τ_r (Fig. 5.2). When the relaxation towards equilibrium

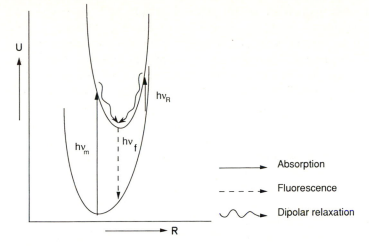

Fig. 5.2. Field diagram describing free energy (U) of solvate in ground and excited state as function of reactive field (R)

is much faster than the average fluorescence lifetime (τ_f), emission will predominantly take place from the equilibrium excited state. The steady-state fluorescence spectrum will then be independent of excitation frequency. This situation is typically encountered in rapidly fluctuating fluorophore environments such as exposed tryptophan residues in proteins. In the reverse situation, where the dipolar relaxation time is much longer than the excited-state lifetime, emission takes place from the energy levels of the non-equilibrium solvate. In that case the red-edge excited emission spectrum will be shifted longwave relative to the main-band excited spectrum [36, 37]. In the intermediate situation ($\tau_f \approx \tau_r$), the relation between the mean wavenumber of the steady-state emission spectrum ($\bar{\nu}$), dipolar relaxation time and average fluorescence lifetime is [38]:

$$\frac{\bar{\nu} - \bar{\nu}_\infty}{\bar{\nu}_0 - \bar{\nu}_\infty} = \frac{\Delta\nu}{\Delta\nu_0} = \frac{\tau_r}{\tau_r + \tau_f}, \tag{1}$$

where $\bar{\nu}_0$ and $\bar{\nu}_\infty$ are the mean frequencies of the unrelaxed ($\tau_r \gg \tau_f$) and completely relaxed ($\tau_r \ll \tau_f$) spectrum, respectively. The relaxed spectrum, typically observed at elevated temperatures, is difficult to monitor in proteins due to temperature denaturation and loss of tertiary structure. A useful expression was obtained which does not contain $\bar{\nu}_\infty$ by assuming: (i) that a unique equilibrium excited state exists independent of excitation wavelength and (ii) that the mean fluorescence lifetime and dipolar relaxation time are invariant to excitation energy [30, 35, 39, 40]:

$$\frac{\bar{\nu} - \bar{\nu}^{\text{edge}}}{\bar{\nu}_0 - \bar{\nu}_0^{\text{edge}}} = \frac{\Delta\nu^{\text{edge}}}{\Delta\nu_0^{\text{edge}}} = \frac{\tau_r}{\tau_r + \tau_f}, \tag{2}$$

where $\bar{\nu}^{\text{edge}}$ is the mean wavenumber of the spectrum obtained upon red-edge excitation. From this expression the dipolar relaxation time in biological molecules can be estimated when the difference in mean frequencies of the unrelaxed spectra $(\bar{v}_0 - \bar{v}_0^{edge})$ is known. Unrelaxed spectra of chromophores in proteins are obtained at low temperatures in cryogenic solvents. With this spectroscopic technique in combination with fluorescence lifetime measurements one can probe the dynamics of the fluorophore environment in a protein [41, 42].

3.2 Fluorescence Lifetime Distributions

The excited-state lifetime of a fluorophore in a protein is extremely sensitive to the physical properties of its environment. A multitude of processes such as collisional quenching, thermal deactivation or energy transfer, causes radiationless decay resulting in a shortening of the lifetime. Under physiological conditions proteins exist in multiple conformational substates (CS) [43, 44]. The population of the free energy levels of CS obeys Boltzmann's distribution law. The rate of interconversion between the states is determined by the activation barrier between them. Each conformation has a different local fluorophore environment resulting in an altered fluorescence lifetime. When the rate of interconversion between CS is much larger than the fluorescence rate of each state, an average fluorescence lifetime will be observed. In the slow exchange limit, each substate will give rise to its own characteristic fluorescence lifetime resulting in a spectrum of decay times. Frauenfelder and coworkers [43, 44] classified the conformational substates in a hierarchical system depending on the activation barrier between the states (abbreviated CS^1, CS^2 and CS^3). Exchange between CS^1 corresponds to conformational transitions involving the whole protein molecule and its hydration layer (activation energies in the order of 100 kJ/mol). CS^2 is a subgroup of CS^1, and exchange between them corresponds to protein domain movements (maximum barrier about 50 kJ/mol). CS^3 transitions pertain to amino acid residue fluctuations (barrier less than 10 kJ/mol). In a protein at room temperature the exchange between states of CS^3 is rapid on a fluorescence time scale. Heterogeneity in the fluorescence decay then originates from degeneracy of CS^1 or CS^2. In order to observe CS^3, fluorescence must be studied in proteins dissolved in cryogenic solvents at low temperatures. In the light of this concept the fluorescence decay pattern of fluorophores in proteins has to be studied over a wide range of temperatures. Lifetime distributions instead of a limited set of discrete exponentials form now the preferential model to analyze the fluorescence decay. Two approaches can be followed. One can define a priori an analytical form of the distribution (e.g. uniform, Gaussian, Lorentzian, unimodal, bimodal). Rejection or acceptance of the model is then based on statistical fit criteria [45–47]. In the non a priori approach, the fluorescence decay is fitted to an exponential series (inverse Laplace transform) with evenly spaced lifetime components (τ) on a $\log \tau$-axis. The optimal spectrum of decay times is recovered by maximizing the

Shannon–Jaynes entropy and minimizing the χ^2 statistics [48]. By using the maximum entropy method (MEM) a solution is chosen with no more correlation than inherent in the experimental data [48]. MEM is the preferable approach in analyzing time-resolved fluorescence in proteins considering the complexity of a multitude of CS.

3.3 Fluorescence Anisotropy

Time-resolved polarized fluorescence spectroscopy is a technique to monitor angular displacements of emission transition moments of fluorescent molecules. It has been applied to protein systems to investigate, among others, rapid angular fluctuations and protein hydrodynamics.

The experimental observable is the fluorescence anisotropy defined as [49]:

$$r(t) = \frac{I_{\parallel}(t) - I_{\perp}(t)}{I_{\parallel}(t) + 2I_{\perp}(t)}, \tag{3}$$

where $I_{\parallel}(t)$ and $I_{\perp}(t)$ are the observed time-dependent parallel and perpendicular polarized components relative to the polarization of the exciting beam. The vertically polarized excitation initially photoselects an anisotropic distribution of fluorophores from the macroscopic isotropic ensemble. This distribution has the form of a second order Legendre polynomial ($P_2(x)$). The information contained in the fundamental anisotropy is lost by several time-dependent mechanisms (vide infra). The general expression relating the time-dependent correlation function of the molecular transition moments with the experimental anisotropy is [50, 51]:

$$r(t) = \langle P_2[\vec{\mu}_a(0) \cdot \vec{\mu}_e(t)] \rangle, \tag{4}$$

where $\vec{\mu}_a(0)$ and $\vec{\mu}_e(t)$ are unit vectors along the transition moment of absorption at time $t = 0$ and emission at time t after excitation, respectively. In a dimeric flavoprotein there are three possible contributions to the depolarization of the fluorescence: rotational diffusion of the whole protein, restricted reorientational motion of FAD and interflavin energy transfer.

3.3.1 Rotation of the Protein

When there is a rigid solution of the protein with non-interacting FAD moieties, the correlation function of the emission transition moments becomes time-independent. One then recovers the fundamental anisotropy containing information about the angle between absorption and emission transition moments (δ):

$$r(t) = r(0) = \tfrac{2}{5} P_2(\cos \delta). \tag{5}$$

The correlation function is time-dependent for a protein rotating in solution. The anisotropy is given by a sum of five exponentials for a completely asymmetric body, which reduces to a triple exponential for a cylindrically symmetric protein [52]. In most experimental cases it is not possible to determine the three individual components by analysis of the fluorescence anisotropy. Instead, a single correlation time is often measured which is the harmonic mean of the three expected values [53].

3.3.2 Restricted Reorientational Motion

When the fluorescent molecule exhibits independent flexibility inside the protein, Eq. (4) can be factorized in three terms [51]:

$$r(t) = \tfrac{2}{5} P_2(\cos \delta) \cdot C_p(t) \cdot C_r(t), \tag{6}$$

where $C_p(t)$ and $C_r(t)$ are the correlation functions of protein tumbling and restricted motion, respectively. $C_r(t)$ is an infinite sum of exponentially decaying functions and cannot be described in close form [50]. Several approximations are described in the literature imposing exact solutions for the derivative at $t = 0$ or the area under the decay [54–58]. All these expressions have the following functional form in common:

$$C_r(t) = \sum_{i=1}^{N} \beta_i \exp(-t/\phi_i) + r_\infty, \tag{7}$$

where the ϕ_i's are the correlation times associated with restricted reorientational dynamics. The value of N is determined by the selected model (typically, N is 1 or 3). It is reasonable to expect that the correlation times will be shorter at higher temperature as relaxation towards equilibrium (r_∞) is more rapidly attained.

3.3.3 Homo-Energy Transfer

The other time-dependent cause for fluorescence depolarization is intersubunit energy transfer between the flavins. This special case of energy transfer between identical chromophores, termed homo-transfer, cannot be observed in the total fluorescence decay. Instead, it is observable in the anisotropy decay when the direction of the transition moments changes during energy transfer [31, 59]. We consider two chromophores with a defined fixed orientation, weakly interacting by dipole–dipole coupling through the Förster mechanism. This system is analogous to a two-state jump model for which the forward and reverse rates of interconversion are identical. It can be shown that the anisotropy decay has the following form [51, 60, 61]:

$$r(t) = \beta_1 \exp(-2k_T t) + \beta_2, \tag{8}$$

where the β_i's are functions of the intramolecular and intermolecular angles between absorption and emission transition moments [62]. The rate of energy transfer k_T is related to geometrical and spectral parameters by the Förster equation [63]:

$$k_T = 8.71 \times 10^{23} \, R^{-6} \, \kappa^2 \, n^{-4} \, k_r J, \tag{9}$$

where R is the intermolecular distance between donor and acceptor, κ is the orientation factor [64] describing the relative orientation of donor and acceptor transition dipoles, n the refractive index of the intervening medium, k_r is the radiative rate of the donor and J is the overlap integral given by:

$$J = \frac{\int\limits_0^\infty \dfrac{F_d(v)\,\varepsilon_a(v)}{v^4} \, dv}{\int\limits_0^\infty F_d(v) \, dv} \tag{10}$$

$F_d(v)$ is the fluorescence intensity of the donor at frequency v and ε_a is the extinction coefficient of the acceptor at v.

The two-state model of energy transfer is valid for a system with uniform transition energies of donor and acceptor molecules. In an inhomogeneously broadened system the transition energies are distributed (Fig. 5.2). This will have consequences for the temperature- and excitation wavelength dependence of the observed rate of transfer. In an inhomogeneously broadened system with $\tau_r \gg \tau_f$, the experimental overlap integral is a static average of the individual overlap integrals of the different dimers in a certain energy configuration. The rate of intersubunit energy transfer is then a distributed function. Upon main-band excitation dimers containing fluorophores with the most probable transition energy are preferentially photoselected. This is equivalent to dimers having the same transition energies for donor and acceptor molecules. The majority of excited molecules will then behave as a homogeneous system. Equation (8) is then a reasonable approximation for the time-dependent anisotropy. Upon red-edge excitation a population of fluorophores is photoselected which have transition energies much smaller than the average one. The occurrence of fluorophores with this small transition energy is infrequent. A fraction of these excited molecules is then not able to transfer energy to an acceptor since the most probable transition energy is larger. Nonetheless, there is a finite chance that a molecule is in a proper energy configuration to act as acceptor. Both fractional populations (transferring and nontransferring) can be determined by fitting edge- and main-band excited anisotropy decays to Eq. (8). The ratio of the preexponential amplitudes (β_1) in both cases yields the fractional population of transferring molecules [10]. Energy transfer in the edge-excited dimers is less efficient than transfer in the main-band excited ones because of the smaller overlap integral (Fig. 3). This will result in a longer correlation time of transfer ($\phi_T = 1/(2k_T)$). Both the red shift of the emission spectrum and the decrease of the transferring fraction upon edge excitation give rise to the Weber red-edge failure of energy transfer [65]. In a dynamic inhomogeneously broadened system ($\tau_r \ll \tau_f$) all energy configurations are sampled during energy transfer

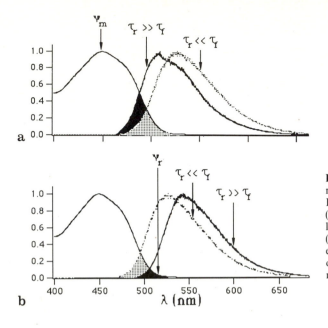

Fig. 5.3a, b. Effect of dipolar relaxation on overlap integral. Panel **a** main-band excitation (v_m), spectral overlap in the limit of slow ($\tau_r \gg \tau_f$) and rapid ($\tau_r \ll \tau_f$) relaxation. Panel **b** red-edge excitation (v_r), spectral overlap in the limit of slow and rapid relaxation

and the observed overlap integral (and thus k_T) is a dynamic average. [66]. The system behaves as if a single energy configuration is present in all dimers. The rate of transfer k_T has then a discrete value independent of excitation energy and the model described by Eq. (8) is exact. The transfer rate in this dynamic regime is smaller than in the static regime because in the latter case the overlap integral is larger (Fig. 5.3). Upon increase of temperature the transfer correlation time (ϕ_T) will become longer. This temperature behavior is opposite to that of reorientational dynamics providing a diagnostic tool to distinguish between both sources of depolarization.

4 Examples

In Fig. 5.4 the time-resolved fluorescence and fluorescence anisotropy profiles are shown for LipdH-AV and GR in aqueous solution at room temperature. It is clear that both types of decays for the two enzymes are entirely different. The spectra of decay times recovered after analysis with MEM are presented in Fig. 5.5. A complex spectrum with five discernible lifetime distributions is recovered for GR. The predominant contribution arises from a peak located at around 30 ps which is probably due to an excited-state complex with Tyr197. The other lifetime components can be ascribed to conformations where the flavin is unfavorably oriented for electron transfer. Two major peaks in addition to a small peak at short lifetimes can be observed in the lifetime spectrum of LipdH-AV. Temperature-dependent studies on the peak positions revealed that

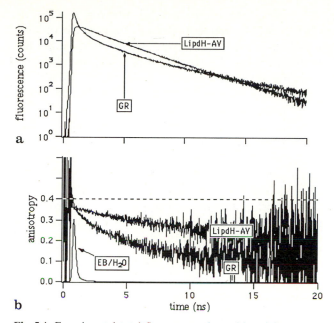

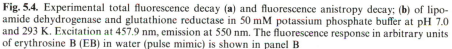

Fig. 5.4. Experimental total fluorescence decay (**a**) and fluorescence anistropy decay; (**b**) of lipo-amide dehydrogenase and glutathione reductase in 50 mM potassium phosphate buffer at pH 7.0 and 293 K. Excitation at 457.9 nm, emission at 550 nm. The fluorescence response in arbitrary units of erythrosine B (EB) in water (pulse mimic) is shown in panel B

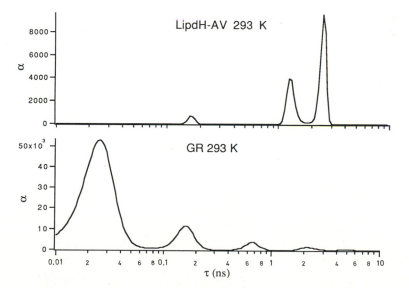

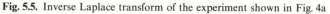

Fig. 5.5. Inverse Laplace transform of the experiment shown in Fig. 4a

rapid exchange between CS takes place in LipdH-AV whereas in GR the interconversion between CS is slow. The activation barrier between the observable states in GR is then of such magnitude that we have to classify them as CS[1]. The transition between states involves the whole protein and the different lifetimes reflect considerable differences in protein conformation. Since the exchange between states is slow, the integrated amplitudes of the individual distributions correspond to the fractional population of the CS [67]. The observed rapid interconversion in LipdH-AV indicates that local protein domains take part in these transitions (exchange between CS[2]).

The fluorescence decay of LipdH-AV was studied in a cryogenic solvent (80% glycerol) in the temperature range 203–303 K. In Fig. 6 the fluorescence decay patterns are shown together with the inverse Laplace transforms. Upon lowering the temperature the shortest lifetime component vanishes from the spectrum and an almost unimodal distribution is recovered. This implies that a

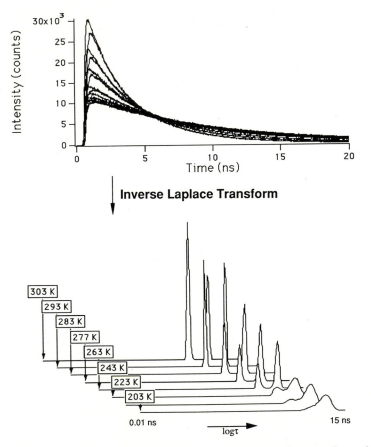

Fig. 5.6. Experimental fluorescence decay and inverse Laplace transform of lipoamide dehydrogenase in 80% glycerol, 50 mM KPi, pH 7.0 at different temperatures

conformational substate at higher enthalpy is depopulated. Since the barycenters [68] of the lifetime classes did not converge in this temperature range, the exchange between CS is slow contrary to the protein in aqueous solution. The activation barrier between states is thus influenced by the more viscous solvent which is analogous to a damping of protein dynamics. It is also noticeable that the shortest lifetime component (as observed in aqueous solution) is not present in the spectrum. This CS must possess the highest altitude in the energy landscape. From a van t'Hoff-plot of the integrated amplitudes the enthalpy difference between the two observed states in 80% glycerol was determined to be 12 kJ/mol. The half width of the lifetime distributions (FWHM) increases at lower temperature. Since the total amount of counts was identical in all experiments, the noise should be comparable allowing for a physical explanation of the relative widths. The empirical activation barrier determined from the FWHM on Arrhenius coordinates is in the order of 4 kJ/mol. A similar activation energy of thermal quenching could be determined from the positions of the barycenters. This is taken to indicate that transitions between CS^3 are observed which are also responsible for collisional fluorescence quenching.

Figure 5.7 gives the distribution of correlation-times between 0.2 and 20 ns as recovered by analysis of the polarized fluorescence decay components of LipdH-AV and GR in aqueous solution at room temperature. For clarity the limiting anisotropy at longer correlation times arising from protein rotation is omitted. Upon main-band excitation the spectrum of LipdH-AV is unimodal whereas the spectrum of GR is bimodal. The unimodal peak for LipdH-AV and the main peak for GR are temperature invariant [69]. This depolarizing process then originates from intersubunit energy transfer between the flavins. The minor peak

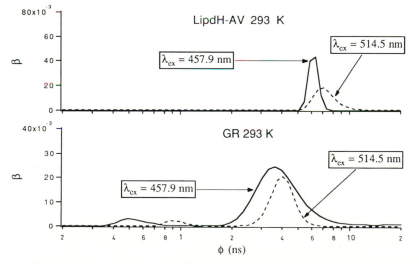

Fig. 5.7. Correlation-time spectrum of lipoamide dehydrogenase and glutathione reductase upon main-band (457.9 nm) and red-edge (514.5 nm) excitation. For clarity only the region between 0.2 and 20 ns is shown

in GR located at shorter correlation times was shown to shift to longer times at decreasing temperature [69]. Apparently, this process must be ascribed to rapid flavin reorientation. Upon red-edge excitation, the integrated amplitude and barycenter of the correlation time distribution are only marginally altered for LipdH-AV. Rapid dipolar relaxation thus takes place and energy transfer occurs from the relaxed excited state. However, the integrated amplitude of the major peak in GR drops to about 50% upon edge excitation. The average fluorescence lifetime of FAD in GR is much shorter than in LipdH-AV (Fig. 5.4a). Relaxation of the solvate is then not complete and about 50% of protein molecules are not transferring energy. From the integrated amplitudes of the correlation time distribution (β_1, β_2) the relative orientation of the transition moments in the two isoalloxazine rings were determined for LipdH-AV ($139° \pm 2°$). The interflavin distance is estimated to be 36 Å from the barycenter of the unimodal distribution [69]. Excellent agreement exists with crystallographic data ($138°$ and 38.6 Å, respectively) [25]. In GR the maximum amplitude of flavin reorientation corresponds to $22° \pm 7°$. The angle between the symmetry axes of the potential wells (in which the flavins reorient) is obtained from the amplitude of the major correlation time component ($109° \pm 7°$) [69].

5 Conclusions

From the set of fluorescence relaxation experiments described, information on protein conformational dynamics can be obtained. We have also demonstrated that the validity of physical models on energy transfer in relation to dipolar relaxation can be tested. The described experimental and analytical methods have broad applicability to other biological systems containing identical fluorophores.

6 References

1. Müller F (1983) Top Curr Chem 108: 71
2. Beechem JM, Brand L (1985) Ann Rev Biochem 54: 43
3. Burstein EA, Vedenkina NS, Ivkova MN (1973) Photochem Photobiol 18: 263
4. Creed D (1984) Photochem Photobiol 39: 537
5. Valeur B, Weber G (1977) Photochem Photobiol 25: 441
6. Cross AJ, Waldeck DH, Fleming GR (1983) J Chem Phys 78: 6455
7. Ruggiero AJ, Todd DC, Fleming GR (1990) J Am Chem Soc 112: 1003
8. Penzer GR, Radda GK (1967) Quart Rev Chem Soc 21: 43
9. Johansson LB-Å, Davidsson Å, Lindblom G, Razi Naqvi K (1979) Biochemistry 18: 4249
10. Bastiaens PIH, van Hoek A, Benen JAE, Brochon JC, Visser AJWG (1992) Biophys (in press)
11. Weber G (1950) Biochem J 47: 114
12. Visser AJWG (1984) Photochem Photobiol 40: 703
13. Visser AJWG, van Hoek A, Kulinski T, LeGall J (1987) FEBS Lett 224: 406
14. van Hoek A, Vos K, Visser AJWG (1987) IEEE J Quantum Electron QE-23: 1812
15. Worthington DJ, Rosemeyer MA (1975) Eur J Biochem 60: 459
16. Worthington DJ, Rosemeyer MA (1976) Eur J Biochem 60: 231
17. Nakashima K, Miwa S, Yamauchi K (1976) Biochim Biophys Acta 445: 309
18. Schulz GE, Zappa H, Worthington DJ, Rosemeyer MA (1975) FEBS Lett 54: 86
19. Williams CH Jr (1976) The Enzymes 13: 89

20. de Kok A, Visser AJWG (1984) In: Bray RC, Engel PC, Mayhew SG (eds) Flavins and flavoproteins. Walter de Gruyter Berlin p 149
21. de Kok A, Bosma HJ, Westphal AH, Veeger C (1988) In: Schellenberger A, Schowen RL (eds) Thiamine pyrophosphate biochemistry, 2 19 CRC Press Boca Raton FL
22. Reed LJ (1974) Acc Chem Res 7: 40
23. Karplus AP, Schulz GE (1987) J Mol Biol 195: 701
24. Schierbeek AJ, Swarte MBA, Dijkstra BW, Vriend G, Read RJ, Hol WGJ, Drenth J, Betzel C (1989) J Mol Biol 206: 365
25. Mattevi A, Schierbeek JA, Hol WGJ (1991) J Mol Biol 220: 975
26. Westphal AH, de Kok A (1988) Eur J Biochem 172: 299
27. Schulze E, Benen JAE, Westphal AH, de Kok A (1991) Eur J Biochem 200: 29
28. Frauenfelder H (1985) In: Clementi E, Chin S (eds) Structure and motion: Membranes nucleic acids and proteins Plenum New York p 169
29. Nemkovich NA, Rubinov AN, Tomin VI (1991) In: Lakowicz JR (ed) Topics in fluorescence spectroscopy, vol 2 Plenum New York p 367
30. Demchenko AP (1986) Essays Biochem 22: 120
31. Bastiaens PIH, Bonants PJM, Müller F, Visser AJWG (1989) Biochemistry 28: 8416
32. de Kok A, Visser AJWG (1987) FEBS Lett 218: 135
33. Kalman B, Sandström A, Johansson LB-Å, Lindskog S (1991) Biochemistry 30: 111
34. Tomin VI, Rubinov AN (1981) J Appl Spectrosc (USSR) 35: 237
35. Demchenko AP (1987) Ultraviolet spectroscopy of proteins. Springer Berlin Heidelberg New York p 183
36. Galley WC, Purkey RM (1970) Proc Natl Acad Sci USA 67: 1116
37. Rubinov AN, Tomin VI (1970) Optics Spectrosc (USSR) 29: 1082
38. Mazurenko YT, Bakhshiev NG (1970) Optics Spectrosc (USSR) 28: 905
39. Demchenko AP (1982) Biophys Chem 15: 101
40. Demchenko AP (1985) FEBS Lett, 182: 99
41. Demchenko AP (1988) Eur Biophys J 16: 121
42. Demchenko AP, Ladokhin AS (1988) Biochim Biophys Acta 955: 352
43. Frauenfelder H, Gratton E (1986) Methods Enzymol 127: 207
44. Frauenfelder H, Parak F, Young RD (1988) Ann Rev Biophys Biophys Chem 17: 451
45. Alcala RJ, Gratton E, Prendergast FG (1987) Biophys J 51: 587
46. Alcala RJ, Gratton E, Prendergast FG (1987) Biophys J 51: 597
47. Alcala RJ, Gratton E, Prendergast G (1987) Biophys J 51: 925
48. Livesey AK, Brochon JC (1987) Biophys J 52: 693
49. Jablonski A (1960) Bull Acad Pol Sci Ser Sci Math Astrom Phys 8: 259
50. Zannoni C (1981) Mol Phys 42: 1303
51. Szabo A (1984) J Chem Phys 81: 150
52. Small EW, Isenberg I (1977) Biopolymers 16: 1907
53. Dale RE, Chen LA, Brand L (1977) J Biol Chem 252: 7500
54. Kinosita K, Kawato S, Ikegami A (1977) Biophys J 20: 289
55. Nordio PL, Segre U (1979) In: Luckhurst GR, Gray GW (eds) Molecular physics of liquid crystals 411 Academic Press London
56. Lipari G, Szabo A (1980) Biophys J 30: 489
57. Zannoni C, Arconi A, Cavatorta P (1983) Chem Phys Lipids 32: 179
58. van der Meer W, Pottel H, Herreman W, Ameloot M, Hendrickx H, Schröder H (1984) Biophys J 46: 515
59. Bastiaens PIH, Bonants PJM, van Hoek A, Müller F, Visser AJWG (1988) Proc SPIE 909: 257
60. Tanaka F, Mataga N (1979) Photochem Photobiol 29: 1091
61. Weber G (1989) J Phys Chem 93: 6069
62. Bastiaens PIH, Mayhew SG, O'Nualláin EM, van Hoek A, Visser AJWG (1991) J of Fluorescence 1: 95
63. Förster Th (1948) Ann Physik 2: 55
64. Steinberg IZ (1971) Ann Rev Biochem 40: 83
65. Weber G, Shinitzky M (1970) Proc Natl Acad Sci USA 65: 823
66. Dale RE, Eisinger J, Blumberg WE (1979) Biophys J 26: 161
67. Mérola F, Rigler R, Holmgren A, Brochon JC (1989) Biochemistry 28: 3383
68. Gentin M, Vincent M, Brochon JC, Livesey AK, Cittanova N, Gallay J (1990) Biochemistry 29: 10405
69. Bastiaens PIH, van Hoek A, Wolkers W, Brochon JC, Visser AJWG (1992) Biochemistry (in press)

6. Fluorescence Spectroscopy on Light Scattering Materials

Dieter Oelkrug, Ulrike Mammel, Manfred Brun, Reiner Günther and Stefan Uhl

Institut für Physikalische und Theoretische Chemie der Universität Tübingen, Tübingen, Germany

1 Introduction

Fluorescence and light scattering are independent processes that appear simultaneously in many systems of biological, analytical, and technical interest (e.g. in plants, tissues, polycrystalline pigments, microscopic preparations, printing products with optical brighteners, polymer blends, thin layer chromatograms, heterogeneous absorbents). In the following, some important consequences of multiple scattering on fluorescence will be discussed, considering especially the spatial distribution of the detectable radiation, fluorescence reabsorption and reemission, samples with gradients in the fluorophore concentration inclusive of microscopic structures, distortion of diffuse reflectance spectra by fluorescence, and finally chemical modifications of the fluorophores by interaction with the phase boundaries of micro-heterogeneous media.

2 Spatial Intensity Distribution of Fluorescence and Determination of Fluorescence Yields

Consider an extended multiple scattering sample of thickness d with plane parallel boundaries. A number N of fluorophores is excited in this sample by irradiating the front boundary with intensity I_0. Contrary to transparent media, N depends not only on the absorption coefficient κ of the fluorophore but also on the scattering coefficient σ of the sample, where both quantities are defined per unit length of the propagating light in arbitrary direction inside the sample [1]. After excitation, the intensity F_f of fluorescence is emitted from the front boundary of the sample and the intensity F_b from the back boundary. Again contrary to transparent media, the ratio F_f/F_b is not unity but increases to infinity with increasing layer thickness. Figure 6.1 shows both parts of fluorescence calculated within the two-flux approximation of Kubelka–Munk (KM) [2] where the optical constants are replaced by formal absorption and scattering coefficients $K = 2\kappa$ and $S = 3\sigma/4$ that are defined per unit length in z-direction perpendicular to the boundary plane of the sample. In thin layers, both F_f and F_b are good quantities for fluorescence analysis. The better choice depends primarily on the geometry of the spectrometer, and it will be pointed out in Sect. 4.2 that the transmission mode should be preferred in samples with a

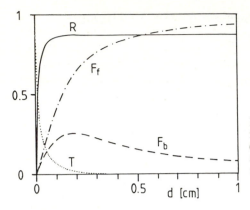

Fig. 6.1. Reflectance (R), transmittance (T), fluorescence in front side (F_f) and back side (F_b) direction of a scattering $(S = 100 \text{ cm}^{-1})$ and absorbing $(K = 1 \text{ cm}^{-1})$ sample as a function of layer thickness d

z-gradient in the concentration of the fluorophore. In thick layers, only the reflectance mode can be used for fluorescence analysis. The angular intensity distribution of F_f now obeys Lambert's cosine-law $f(\vartheta) = f_0 \cos \vartheta$ and the maximum fluorescence intensity, that will be obtained for "infinite" layer thicknesses, is then

$$F^S_\infty = \int_0^{2\pi} \int_0^{\pi/2} f^S_0 \cos \vartheta \sin \vartheta \, d\vartheta \, d\varphi = \pi f^S_0 = I_0 \Phi_F (1 - R_{\infty, \text{ex}}). \tag{1}$$

Here $R_{\infty, \text{ex}}$ is the diffusely reflected part of the incident radiation, and ϕ_F is the fluorescence quantum yield.

In a transparent sample, the corresponding quantities are $f(\vartheta) = f_0 = \text{const}$, and

$$F^T = \int_0^{2\pi} \int_0^{\pi} f^T_0 \sin \vartheta \, d\vartheta \, d\varphi = 4\pi f^T_0 = I_0 \Phi_F (1 - T_{\text{ex}}). \tag{2}$$

Here T_{ex} is the transmitted part of the incident radiation. Usually, only a small region of F is detected that is emitted into a solid angle $d\omega$ perpendicular to the boundary plane. In this case we obtain

$$dF^S = f^S_0 \, d\omega \qquad dF^T = f^T_0 \cdot n^{-2} \, d\omega. \tag{3}$$

The equation on the right side considers the relative refractive index n of the transparent sample relative to the environment (air). If we start in both samples with equal numbers of excited species, equal fluorescence yields and $n = 1.4$–1.5, the relation

$$dF^S \approx 8 \cdot dF^T, \tag{4}$$

is obtained, i.e. the scattering sample emits about eight times more intense than the transparent one. Of course, the fluorescence of the transparent sample is not lost. It is either emitted through the back boundary or guided by total reflection to the small end sides of the sample.

Equation (1) describes the most convenient situation for the determination of luminescence quantum yields in scattering media. The main problem is the preparation of "infinitely" thick layers that are by far not so easy to obtain as for reflectance measurements. This can be seen from Fig. 6.1, where the upper limit of the reflectivity is reached already at layer thicknesses of $d \lesssim 0.1$ cm but the upper limit of F_f not yet even at $d \gtrsim 1$ cm. In practice, the situation is somewhat less dramatic since all scattering substrates show a small background absorption that reduces the penetration depth of fluorescence below 1 cm. All luminescence yields of this paper have been determined with $d = 1$ cm samples using Eq. (1) and taking into account the reabsorption of luminescence according to Sect. 4. In order to avoid absolute measurements, the yields are given relative to perylene adsorbed on microporous silica with $\phi_F = 0.85$ or 3,6-bis-(dimethyl-amino)-10,10-dimethylanthrone-9 adsorbed on microporous silica gel with $\phi_F = 0.95$ [3].

3 Sensitivity and Concentration Dependence of Fluorometric Analysis

The high sensitivity of fluorescence spectroscopy is of course maintained also in strongly scattering media, especially under consideration of the intensity gain mechanism of Eq. (4). The limit of detection is therefore usually not determined by the amount of the fluorophore but by Raman-scattering or self-fluorescence of the substrate. Figure 6.2 shows the total Raman + fluorescence emission of a specially weakly fluorescent cellulose filter paper and the difference spectrum after adsorption of 1 pg fluorescein on the illuminated area. The difference spectrum can be clearly assigned to the fluorescence of fluorescein (see e.g. Fig. 8), but its intensity is already as low as the background emission, and thus the detection limit will be reached at about 300 fg. In all cases of non-minimized background emission the limit will be definitely higher.

In order to obtain quantitative relations between fluorescence intensity and concentration, not only the emission properties of the substrate but also its

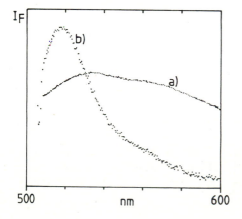

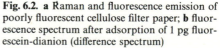

Fig. 6.2. a Raman and fluorescence emission of poorly fluorescent cellulose filter paper; **b** fluorescence spectrum after adsorption of 1 pg fluorescein-dianion (difference spectrum)

absorption properties must be considered. The general Eq. (1) for infinite layer thickness has to be modified as

$$F^S_\infty = I_0 \Phi_F (1 - R_{\infty,\text{ex}}) K_F / K, \tag{5}$$

where $K_F = 2\varepsilon_F \cdot c_F$ is the absorption coefficient of the fluorophore at the excitation wavelength, and K is the sum of the absorption coefficients of all species absorbing at the excitation wavelength including the background K_B of the substrate. In the simplest case we obtain $K = K_F + K_B$. If Eq. (5) is expressed in terms of the Kubelka–Munk function $R_\infty = (K + S)S^{-1} - \langle (K + S)^2 S^{-2} - 1 \rangle^{1/2}$, two limiting relations between fluorescence intensity and fluorophore concentration are obtained:

1. The background absorption is completely negligible and the absorption coefficient of the fluorophore is low, i.e. $K_B \ll K_F \ll S$. Then

$$F_\infty \approx I_0 \cdot \Phi_F (2K_F / S)^{1/2} \sim c_F^{1/2}. \tag{6}$$

The fluorescence intensity increases with the square root of c_F, so that the fluorescence is of considerable intensity even at extremely low fluorophore concentrations. However, this situation is of minor importance in experimental studies. It could be verified [4] e.g. with the fluorophore 1-naphthol adsorbed on freshly prepared magnesium oxide powder (Fig. 6.3) that has an extremely high scattering coefficient of $S \approx 10^3 \text{ cm}^{-1}$ and almost no absorptivity in the visible and near UV range.

2. In a more realistic situation, the background absorption is much higher than the absorption coefficient of the fluorophore, i.e. $K_F \ll K_B$. This is usually true for fluorescence trace analysis, since many substrates possess absorptivities in the order of $K_B = 0.01$–1 cm^{-1}. The reflectivity of the sample is then approximately equal to the reflectivity of the background, R_B, and independent of c_F. Now Eq. (5) can be written as

$$F_\infty \approx I_0 \cdot \Phi_F (1 - R_{\infty,B}) K_F / K_B \sim c_F. \tag{7}$$

The fluorescence is directly proportional to c_F, as in transparent media of low absorptivity. Figure 6.4 shows the example of fluorescein adsorbed on microcrystalline alumina and the corresponding intensity analysis in alkaline

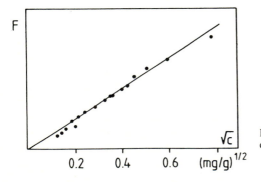

Fig. 6.3. Concentration dependence of fluorescence intensity of 1-naphthol on magnesia

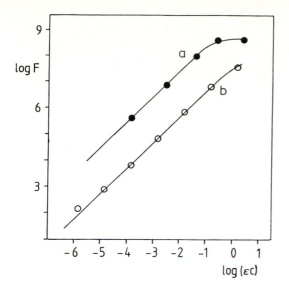

Fig. 6.4a, b. Fluorescence intensity as a function of the fluorophore concentration. **a** fluorescein-dianion adsorbed on microcrystalline alumina; **b** fluorescein-dianion in alkaline aqueous solution

solution. The slope of the $\log F/\log c$ – plot is very close to unity, which proves the validity of Eq. (7).

4 Chemical Interactions Between Fluorophore and Substrate

The fluorescence of many organic molecules is modified in heterogeneous media by the binding forces to the microcrystalline or microporous substrate. These forces may reduce or, sometimes, enhance the molecular mobility of the fluorophore and change its electron distribution by, e.g. intermolecular electron and proton transfer reactions. As an example, the luminescence of adsorbed 5,6-benzoquinoline will be discussed. The results are typical for all mono-aza analogues of naphthalene, phenanthrene or anthracene, and can be subdivided into three main types of interaction:

1. hydrogen-bonding between weakly polar surface hydroxyl groups of e.g. silica gels and the nitrogen atoms of the fluorophores. The fluorescence spectra (see Fig. 6.5, upper), the fluorescence quantum yields ($\phi_F = 0.3$–0.5) as well as the fluorescence decay times are very similar to those in polar solvents like ethanol. In addition, weak room temperature phosphorescence (RTP) with $\phi_P < 0.01$ is observable on silica gel surfaces that have been partly dehydroxylated by preheating them prior to adsorption to temperatures of $T_a \approx 600°C$ [5]. Some complications may arise in the fluorescence properties when the basicity of the excited fluorophore is high enough to form protonated cations (see point 2). Then the fluorescence consists of a dynamic superposition of two types of emissions with wavelength dependent decay characteristics [6].

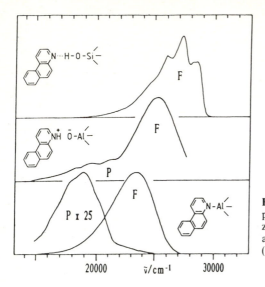

Fig. 6.5. Fluorescence (*F*) and room-temperature phosphorescence (*P*) of 5,6-benzoquinoline on silica ($T_a = 100°C$, top), alumina ($T_a < 100\,°C$, middle) and alumina ($T > 300\,°C$, bottom)

2. protonation at the nitrogen atoms on strongly polar, fully hydroxylated surfaces of e.g. microcrystalline alumina, titania or tin(IV) oxide [6]. The fluorescence spectra of the protonated heterocycles are red-shifted compared to the hydrogen-bonded species and have lost most of their vibrational structure (see Fig. 6.5 middle). The fluorescence yields are high ($\phi_F \approx 0.7$) and also the RTP can achieve considerable yields of $\phi_P \approx 0.05$.

3. polar atomic bonding between the nitrogen lone pair or the π-electron system of the heterocycles and coordinatively unsaturated (CUS) Lewis-acidic metal centers on strongly polar, partly dehydroxylated metal oxides. Most information is available on thermally pretreated, highly distorted γ-alumina that is commonly used for chromatographic purposes. Almost all condensed aromatics show a broad unstructured fluorescence on this surface, as can be seen in Fig. 6.5, lower. In the case of the parent hydrocarbons the fluorescence has been assigned by us to the CT-emission of EDA-complexes with two- or three-center binding of the aromatic π-system to adjacent CUS aluminium atoms [7–9]. Figure 6.6 shows the CT-transition energies of a series of adsorbates versus the first ionization potentials of the free molecules. The results of the hydrocarbons and of the heterocycles give good straight lines with slopes of unity, as it is expected according to Mulliken's charge transfer transition theory [10]. Since the straight line of the heterocycles is displaced by $2500\ \text{cm}^{-1}$ to lower transition energies, we conclude that the donor–acceptor distance is lower and the binding is stronger than in the adsorbed hydrocarbons. Since, in addition, the CT-complexes of the heterocycles are formed at relatively low T_a, where mainly isolated and no adjacent CUS metal atoms are present at the surface [11], we conclude that the heterocycles are adsorbed by one-center N_{donor}–$Al_{acceptor}$ bonds.

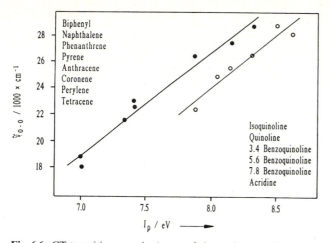

Fig. 6.6. CT-transition energies (mean of absorption and fluorescence maxima) of aromatics (*closed cycles*) and azaaromatics (*open cycles*) adsorbed on alumina versus the π-ionization potentials I_P of the free molecules. The aromatics are ordered from top to bottom in the direction of decreasing I_P

Table 6.1. Room-temperature and low-temperature phosphorescence yields of some rigid luminophores adsorbed on thermally pretreated ($T_a = 300°C$) alumina

Luminophore	ϕ_P adsorbed on alumina		ϕ_P in EPA [15, 16]
	$T = 300$ K	$T = 77$ K	$T = 77$ K
Naphthalene	0.15	0.2	0.04
Phenanthrene	0.35	0.58	0.1
Coronene	0.35[a]	0.12	0.11
5,6-benzoquinoline	0.05	0.06	0.1

[a] inclusive of E-type delayed fluorescence

The fluorescence of the adsorbed aromatics is accompanied very often by intense RTP [12–14]. As can be seen from Table 6.1, the RTP yields of the adsorbates can be very high, and temperature has only a moderate effect on the non-radiative $T_1 \rightarrow S_0$ deactivation. The reasons for this behavior are the absence of mobile molecular oxygen and the rigid binding of the aromatics to the alumina surface, so that the density of states in the preexponential factor of the non-radiative $T_1 \rightarrow S_0$ rate constant is very low. The situation changes completely as soon as the density of states is increased by introducing flexible substituents into the luminophores. The examples of Table 6.2 show absolutely no RTP on alumina or silica. Their phosphorescence yields at low temperatures are mainly determined by the $S_1 \rightarrow T_1$ population rate of the triplets and this rate depends again on the flexibility of the adsorbates. Molecules like Michler's ketone (MK) or its isopropylene bridged derivative (MKB) are rigidly adsorbed on silica gel [3]. Consequently their triplet formation yields are lowered resulting in much higher fluorescence and lower phosphorescence yields than in

Table 6.2. Low-temperature fluorescence and phosphorescence yields of some flexible luminophores adsorbed on silica gel and dissolved in ethanol

Luminophore		$\phi_F (T = 77\ K)$		$\phi_P (T = 77\ K)$	
		Silica gel	Ethanol	Silica gel	Ethanol
	MK	0.25	0.06	0.5	0.8
	MKB	0.95	0.2	0.03	0.6
	DMABN	0.05	0.3	0.65	0.6
	DMAPS	0.15	0.35	0.5	0.55

glassy ethanol. Molecules like DMABN or DMAPS (see Table 6.2) keep their flexibility in the adsorbed state because not all rotating parts are in contact with the surface. The fluorescence yields are now distinctly lower than in glassy ethanol, but the phosphorescence yields are not correspondingly increased since an appreciable proportion of the excited singlets deactivates by non-radiative $S_1 \rightarrow S_0$ internal conversion.

5 Problems of Quantitative Fluorescence Spectroscopy

The photometric accuracy of transparent systems is hardly to achieve in multiple scattering media, but if the limitations are understood, the errors can be minimized.

5.1 Reabsorption and Reemission

The reasons for fluorescence reabsorption are the same as in transparent samples: overlap between the absorption and emission spectra of the fluorophore, and absorption by the substrate. However, the effect is reinforced in scattering media by the long mean path lengths, especially of that part of fluorescence that has been originally emitted into the interior of the sample and has multiply changed its direction before it emerges from the front surface.

The degree of reabsorption has been calculated within the limits of the KM model [17] and also the influence of reemission has been considered [18]. The results show that very low fluorophore concentrations or very low layer thicknesses are necessary in order to avoid reabsorption. Both conditions are demonstrated in Fig. 6.7 for perylene adsorbed on silica gel. The 0-0-transition of the fluorescence band is completely suppressed in thick samples of moderate perylene concentration. The band appears clearly in very thin layers of the same

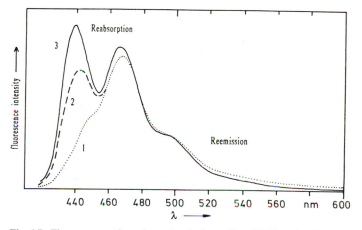

Fig. 6.7. Fluorescence of perylene adsorbed on silica: *1* 0.09 mg/g, thick layer *2* 0.09 mg/g, thin layer 3.5×10^{-4} mg/g thick layer

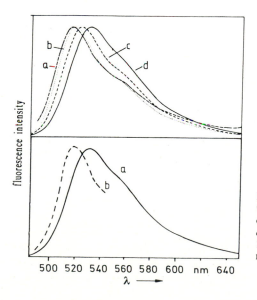

Fig. 6.8. *Top*: Concentration dependence of the fluorescence spectra of fluorescein-dianion on alumina *a* 5 ng/g, *b* 230 ng/g, *c* 1.4 μg/g, *d* 14 μg/g; *bottom*: Emission of 14 μg fluorescein-dianion/g alumina, *a* as measured and, *b* corrected for reabsorption

sample and has gained its original intensity in very low-concentrated samples. The reemission of the selfabsorbed light can also be observed in spectrum (1) that is more intense in the long wavelength region than spectrum (2).

Reabsorption of fluorescence may lead to misassignements of the fluorescent species, especially in cases of bands with no vibrational structure. The fluorescence of e.g. fluorescein changes its position on alumina systematically with the adsorbate concentration, see Fig. 6.8, top, and this might be assigned to the occupation of surface sites of different basicity. However, correction of reabsorption will shift all spectra to the same position, as it is shown in Fig. 6.8, bottom, for the most concentrated sample.

5.2 Inhomogeneous Distribution of the Fluorophores

Most of the experimental and theoretical work on multiple light scattering has been done under the condition of homogeneous irradiation and under the assumption of spatially constant absorption and scattering coefficients. These assumptions must not necessarily be true in laterally or vertically structured materials, as in biological samples, offset printing products or high performance thin layer chromatograms. The assumptions can also become invalid by avoidable or unavoidable imperfections of the sample preparation.

Samples with inhomogeneous depth profiles of the fluorophores:

In non-scattering systems, the optical density of a given amount of molecular absorber is independent of concentration inhomogeneities in the direction of the incident light. This situation can change drastically in multiple scattering systems. The results of a KM model calculation for a two-layer system are shown in Fig. 6.9. A given amount of absorber is compressed first more and more

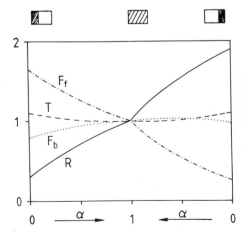

Fig. 6.9. Reflectance (R), transmittance (T), fluorescence in front side (F_f) and back side (F_b) direction of a scattering ($Sd = 1$) and absorbing ($\alpha Kd = 1$) layer of thickness d as a function of the depth distribution of the absorbing fluorophore. The results are normalized to homogeneous distribution ($\alpha = 1$, *middle* of the abscissa). *Left* (right) direction of the abscissa: The absorber is concentrated in the part αd of the layer of the illuminated (unilluminated) side. The part $(1 - \alpha)d$ is free of absorber but still scattering

from the illuminated front side into the direction of the back side and then the other way round. The reflected intensities of the incident and emitted light (R, F_f) deviate extremely from the homogeneous case $(\alpha = 1)$. The slopes of both curves are non-zero at $\alpha = 1$, i.e even small deviations from homogeneity will change the reflectance properties considerably (and may explain many difficulties in quantitative reflectance spectrometry). In contrast to this result, the transmitted intensities (T, F_b) are only little dependent on α, and especially the slopes of both curves are approximately zero at $\alpha = 1$. We therefore recommend measuring, if possible, in the transmission mode.

Lateral micro-structured samples:

Samples with lateral changing absorption and fluorescence properties (e.g. high performance thin layer chromatograms) can be examined with microscopic resolving scanning spectrometers. The absorptivity of those samples is principally not exactly proportional to the geometrical cross sections of the absorbing spots because scattered light diffuses from non-absorbing to absorbing regions and vice versa. The fluorescence of the spots is correspondingly washed out.

The lateral displacements of scattered incident and emitted light were calculated by us with diffusion theory and Monte Carlo methods [19, 20]. One of the results is shown in Fig. 6.10a in an intensity-distance diagram, as it is obtained by a scanning microdensitometer. A second result is fitted in Fig. 10b to one of our experiments, and plotted as radial intensity distribution function $2\pi \cdot r \cdot I(r)$. In all cases a focussed Gaussian laser beam (e^{-2}-width $= 12 \ \mu m$) was used as source of irradiation. Experimental and calculated traces coincide very well, and as can be seen from the figures the reflected light is displaced in cellulose paper by an average of about 80 μm from the origin. All objects up to ten times of this dimension are more or less disturbed in their absorption and fluorescence properties by scattering.

The consequences of this result will be treated in more detail for an absorbing and fluorescent chromatographic spot. The chromatogram is irradiated, as shown in Fig. 6.11, by a homogeneous parallel beam of light of intensity I_0, i.e. under usual macroscopic conditions. The cylindrical spot of radius r_0 is exposed not only to I_0, but in addition to scattered light I_{in} from the non-absorbing regions outside the spot. Some light I_{out} is also leaving the spot volume through the side boundaries, but $I_{out} < I_{in}$. Similar considerations are valid for the fluorescence that is created inside the spot. Some fluorescence flux F_{out} is leaving the side boundaries and some flux F_{in} is reflected back by scattering, where now $F_{out} > F_{in}$. The specific absorptivity of the spot is increasing by the asymmetrical lateral components of the I – fluxes (in the example of Fig. 6.11 the increase is about 15%), and the scanning reflectogram looks as if the spot had gained in magnitude. A similar effect is observed in the scanning fluorogram that spreads over the physical boundaries of the spot (in Fig. 6.11 by about 13%). The absorption gain disappears only for $I_{out} = I_{in}$, i.e.

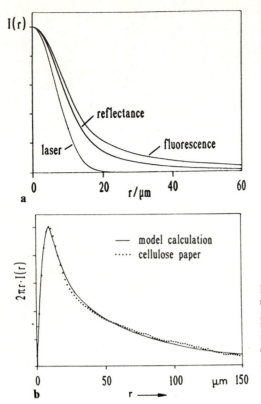

a

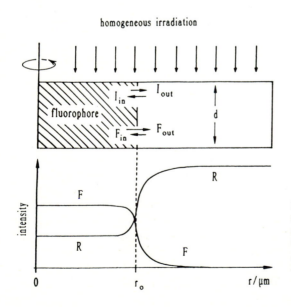

b

Fig. 6.10. **a** Lateral diffusion of light in a scattering layer (d = 200 μm, σ = 100 cm^{-1}, χ = 100 cm^{-1}, e^{-2} = 12.5 μm); **b** radial distribution of the reflectance of cellulose paper under focussed laser irradiation. Model calculation with d = 160 μm, σ = 350 cm^{-1}, χ = 0.1 cm^{-1} and e^{-2} = 12.5 μm

Fig. 6.11. Lateral resolved reflectance and fluorescence intensities of a cylindrical spot. Model calculations with r_0 = 500 μm, d = 200 μm, σ = 300 cm^{-1}, χ = 100 cm^{-1} inside the spot, and χ = 0 on the outside

for vanishing fluorophore concentration in the spot. In this case the absorptivity is of course no longer measurable, but the fluorescence is, and its integrated intensity is now really proportional to the amount of fluorophore, whereas its intensity distribution still spreads over the physical boundaries of the spot.

In a second arrangement only the spot area or part of it is irradiated, (i.e. $I_{out} > I_{in}$ and $F_{out} > F_{in}$. The specific absorptivity and fluorescence intensity are now lower than in the homogeneous case. The error is highest for vanishing absorber concentration in the spot, becomes small for strongly absorbing spots (with the data of Fig. 6.11 the fluorescence intensity is reduced only by 3%), and will be zero for total absorption ($I_{out} \rightarrow 0$). According to these results, the type of irradiation should be primarily chosen under the aspect of spot absorptivity.

5.3 Reflectance Spectroscopy of Fluorescent Samples

Conventional spectroscopic arrangements (light source L – monochromator M – sample S – detector D) will always measure an apparent reflectance $R' = R_{\lambda_1} + bF_{\lambda_2} = R_{\lambda_1} + b(1 - R_{\lambda_1})\phi_F$, that is the sum of the diffuse reflectance at the excitation wavelength λ_1 and the fluorescence at λ_2. Since both types of radiation have the same angular intensity distribution ($b \approx 1$), the difference between R' and R can be very high. The factor b considers also the relative sensitivity of the detection system at λ_2 and λ_1. As soon as $b \cdot \phi_F > 1$, the apparent reflectivity becomes larger than unity and increases when the real reflectivity decreases. Figure 6.12 shows the reflectance spectra of a highly

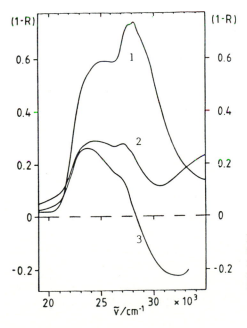

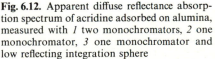

Fig. 6.12. Apparent diffuse reflectance absorption spectrum of acridine adsorbed on alumina, measured with *1* two monochromators, *2* one monochromator, *3* one monochromator and low reflecting integration sphere

fluorescent adsorbate ($\phi_F \approx 0.8$) measured with different spectrometers. Spectra 2 and 3 are obtained with integrating spheres of high and low reflectivity. The factor b increases when the reflectivity of the sphere decreases. This effect appears especially in the UV, and shifts the apparent reflectance to remarkably high values. More accurate spectra are obtained with the arrangement L–S–M–D, that is realized in most diode array spectrometers. The main disadvantages of this arrangement are the high intensities of irradiation and the spectral distortions by second-order reflection of the grating. In order to avoid these effects we recorded spectrum 3 with a more sophisticated spectrometer L–M–S–M–D, that has a very low straylight level and gives the best 'real' absorption spectra of strongly luminescent samples.

Acknowledgement. This work was supported by the Deutsche Forschungsgemeinschaft (Oe 57/12).

6 References

1. Chandrasekhar S (1960) Radiative transfer, Dover Publications, New York
2. a. Kubelka P, Munk F (1931) Z Techn Phys 12: 593 b. Kubelka P (1948) J Opt Soc Am 38: 448
3. Günther R (1991) Ph.D. Thesis, University of Tübingen
4. Kessler RW, Uhl S, Oelkrug D (1981) Le vide et les couches minces 209: 1338
5. Oelkrug D, Uhl S, Gregor M, Lege R, Kelly G, Wilkinson F (1990) J Mol Struct 218: 435
6. Rempfer K, Uhl S, Oelkrug D (1984) J Mol Struct 114: 225
7. Oelkrug D, Radjaipour M (1980) Z Phys Chem NF 123: 163
8. Kessler RW, Uhl S, Honnen W, Oelkrug D (1981) J Luminesc 24/25: 551
9. Honnen W, Krabichler G, Uhl S, Oelkrug D (1983) J Phys Chem 87: 4872
10. a. Mullikan RS, Pearson WB (1972) Annu Rev Phys Chem 13: 107 b. Briegleb G (1961) Elektronen Donator-Acceptor-Komplexe, Springer, Berlin Göttingen Heidelberg
11. Knözinger H, Ratnasamy P (1978) Catal Rev Sci Eng 17: 31
12. a. Schulman EM, Walling Ch (1973) J Phys Chem 77: 902 b. Schulman EM, Parker RT (1977) J Phys Chem 81: 1932
13. Oelkrug D, Plauschinat M, Kessler RW (1979) J Luminesc 18/19: 434
14. Richmond MD, Hurtubise RJ (1991) Anal Chem 63: 169, 1073
15. Li R, Lim EC (1972) J Chem Phys 57: 605
16. Masetti F, Mazzucato U, Birks JB (1975) Chem Phys 9: 301
17. Oelkrug D, Kortüm G (1968) Z Phys Chem NF 58: 181
18. Gade R, Kaden U (1990) J Chem Soc Faraday Trans 86(22): 3707
19. Ishimaru A (1978) Wave propagation and scattering in random media, vol. 1, Academic New York
20. Cashwell ED, Everett CJ (1959) Monte Carlo method for random walk problems, Pergamon, London

7. Optical Sensors Based on Fluorescence Quenching

Wolfgang Trettnak

Institute for Optical Sensors, Joanneum Research, Steyrergasse 17, A-8010 Graz, Austria

1 Introduction

A sensor is a device capable of continuously monitoring a physical parameter or the concentration of an analyte. Among the various types of sensors, electrochemical and optical sensors form the two largest groups. Optical sensors ("optrodes" or "optodes") are mainly based on the detection of changes in absorbance, reflectance, fluorescence or chemiluminescence. But also Raman-scattering, refractive index, light polarization, light scattering and other optical properties have been used as analytical parameters.

At the moment, optical sensors based on fluorescence spectroscopy are the most highly developed, because of the higher sensitivity, selectivity and versatility of fluorescence spectroscopy when compared to absorption spectroscopy. Measuring chemiluminescence seems to be very attractive too, because the light is generated through chemical or biochemical reactions. No light source is needed and the experimental set-up can be kept very simple. Unfortunately, the number of applications for this kind of sensor is rather limited at the moment.

Fluorescence quenching refers to any process which decreases the fluorescence intensity of a certain fluorophore. A variety of processes can result in such a decrease. Some examples are collisional or dynamic quenching, complex formation, excited state reactions and energy transfer. However, this report will deal solely with optical sensors based on dynamic (collisional) quenching and static quenching (through complex formation).

2 Some Principles of Fluorescence Quenching

Dynamic fluorescence quenching results from collisions between the fluorophore in its excited state and the quenching molecule. The fluorophore returns to the ground state without emission of light. In the case of static fluorescence quenching, a non-fluorescent complex is formed between the fluorophore and the quencher. Usually, only a fluorophore which is not complexed can exhibit fluorescence. In both cases, the fluorescence intensity measured is related to the concentration of the quencher. Therefore, the quenched fluorophore may serve as an indicator for the quenching agent.

In many cases, dynamic fluorescence quenching can be described by the Stern–Volmer equation:

$$I_0/I = \tau_0/\tau = 1 + K_d \cdot [Q] = 1 + k_q \cdot \tau_0 \cdot [Q], \qquad (1)$$

I_0 and I are the fluorescence intensities in absence and in presence of a quencher, respectively, K_d is the quenching constant or Stern–Volmer constant, $[Q]$ is the quencher concentration and k_q is the bimolecular quenching constant. τ_0 and τ are the lifetimes of the excited state of the fluorophore in absence and in presence of a quencher, respectively. Since the collision of the fluorophore with the quencher occurs in its excited state, the lifetime of the excited state is reduced too. Dynamic fluorescence quenching is a diffusion dependent process and therefore is also influenced by solvent viscosity and temperature. The quenching efficiency increases with increasing temperature (Fig. 7.1).

Static quenching can be described by an equation similar to Eq. (1):

$$I_0/I = 1 + K_s \cdot [Q]. \qquad (2)$$

The quenching constant K_s is now identical with the association constant of the complex formed between fluorophore and quencher. In the case of static quenching a fraction of the fluorophore is removed by complexation whereas the fluorescence of the uncomplexed portion remains unperturbed. Therefore the lifetime of the excited state of the fluorophore is unchanged. Since an increase in temperature results in a decrease in the stability of the complex, the effect of static quenching is lower at higher temperature (Fig. 7.1). A ground state

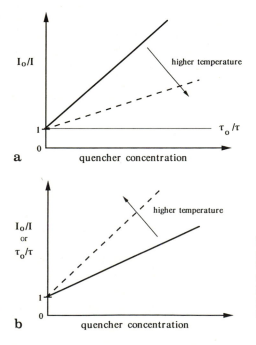

Fig. 7.1. Comparison of **a** static and **b** dynamic fluorescence quenching. I_0 and I are the fluorescence intensities in absence and in presence of a quencher, whereas τ_0 and τ are the lifetimes of the excited states of the fluorophore in absence and in presence of a quencher, respectively

complex formation also frequently results in a change in the absorption spectrum of the fluorophore, whereas collisional quenching only affects the excited state. A comparison of dynamic and static fluorescence quenching is shown in Fig. 7.1.

Equations (1) and (2) are valid only for a single quenching process. Frequently, non-linear relationships are obtained between I_0/I and the quencher concentration, which is very often due to more than one quenching process occuring at the same time. In simple cases of combined dynamic and static quenching it is possible to separate the contributions of both quenching processes mathematically. In addition, the portion of dynamic quenching may also be determined by lifetime measurements [1].

3 Oxygen Sensors

Molecular oxygen is one of the best known collisional quenchers. It quenches the fluorescence of a great number of fluorophores. Among these are polycyclic aromatic hydrocarbons, metal-organic complexes of ruthenium, osmium, iridium and platinum, and a variety of surface-adsorbed heterocyclic compounds including acridine yellow, fluorescein, rhodamin B, tetraphenylporphyrin and perylene tetracarboxylic acid N-alkylimide [2].

Among the group of the polycyclic aromatic hydrocarbons especially decacyclene shows some properties, which make it very suitable for constructing an optical sensor. It shows a high quantum yield, high photostability and can be excited with visible light [3]. Solubilized decacyclene (decacyclene which has been made polymer-soluble by butylation) can be excited even at 450 nm whereas most of the other polycyclic aromatic hydrocarbons require UV light. Its fluorescence can be measured approximately from 470 to 550 nm. The indicator is typically dissolved in a thin layer of silicone which is attached, for example, to the end of a fiber optic bundle (Fig. 7.2) [3]. Silicone is used for the preparation of oxygen-sensitive layers because it shows a high solubility and permeability for oxygen. Figure 7.3 shows the calibration graph and the Stern–Volmer plot of an oxygen sensor based on solubilized decacyclene. The response time for gases varies from a few seconds up to 15 s (depending on the thickness of the layer) for 90% of the total signal change and increases up to several minutes in liquid samples. It was demonstrated that it is even possible to integrate decacyclene in a Langmuir–Blodgett layer consisting of decacyclene and octadecanol in a ratio of 1 : 10. This lipid layer also showed a good response towards oxygen, but needs a more sophisticated optical arrangement in order to achieve a high signal-to-noise ratio [4]

A good representative of the metal-organic indicator group is the ruthenium(II) complex of bipyridyl. This type of fluorophor combines a large Stokes' shift with good photostability and high sensitivity towards oxygen. It shows excitation and emission maxima of approximately 470 nm and 610 nm, respectively. It is typically adsorbed on silica gel beads which are embedded in a

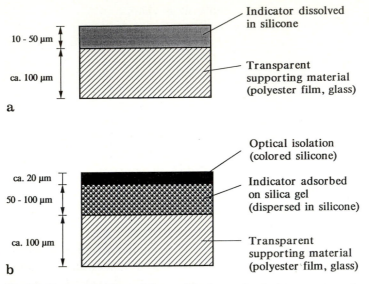

Fig. 7.2. Cross-section through the sensitive layers of optical oxygen sensors based on **a** solubilized decacyclene and **b** ruthenium(II)tris(bipyridyl) as the oxygen indicators

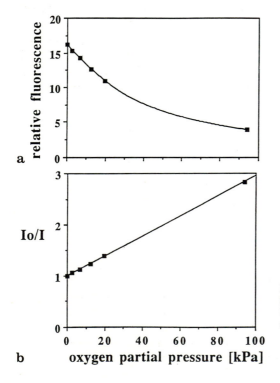

Fig. 7.3. a Calibration graph and **b** Stern-Volmer plot of an oxygen sensor based on solubilized decacylene in silicone

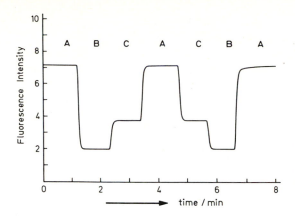

Fig. 7.4. Response of an oxygen sensor based on ruthenium(II)-tris(bipyridyl) towards nitrogen (*A*), oxygen (*B*) and air (*C*)

silicone matrix [5]. The silicone/silicagel layer is again covered with a colored silicone to prevent ambient light and sample fluorescence to enter the light guide (Fig. 7.2). The response of a ruthenium(II)-tris(bipyridyl) based sensor towards nitrogen, air and oxygen is shown in Fig. 7.4. Its response times are in the order of 30–40 s for reaching 99% of its final value in the case of gases. The same sensor has shown to be suitable for measuring the lifetime of the excited state instead the fluorescence intensity. Measuring fluorescence lifetimes reduces the problems of photobleaching and wash-out of the dye and from fluctuations in the intensity of the light source [6].

Despite the good results obtained with optical oxygen sensors it is necessary to mention that fluorescence quenching is usually not a very selective process. Very often, a variety of substances may act as quenchers of a single fluorophore. However, this fact also may lead to the development of novel sensors for new analytes.

4 Halothane Sensor

The fluorescence of decacyclene is not only quenched by molecular oxygen, but also by aryl and alkyl halides, especially by the iodo- and bromo-compounds. This fact was utilized for the determination of halothane (2-bromo-2-chloro-1,1,1-trifluoroethane), which is an inhalation narcotic frequently used in anesthesia [3]. Halothane very efficiently quenches the decacyclene fluorescence, but has to be determined by a two-sensor method because of the interference of oxygen. One sensor measures halothane and oxygen, whereas a second sensor is made halothan-insensitive by covering it with a teflon membrane. The two parameters can be measured simultaneously. Halothane was determined in nitrogen, air and pure oxygen in a concentration range of 0 to 4% by volume. The response times were 15–20 s for 90% of the total signal change. Carbon monoxide, dinitrogen monoxide or fluorans were found not to interfere.

5 Sulphur Dioxide Sensor

Sulphur dioxide is an efficient quencher of the fluorescence of certain polycyclic aromatic hydrocarbons, such as fluoranthene, benzo(b)fluoranthene or pyrene [7]. A fiber optic fluorosensor based on benzo(b)fluoranthene was constructed by dissolving the indicator dye in a thin silicone layer [8]. It was excited at 330 nm and its emission was measured at 450 nm. The sensor showed a detection limit of about 0.01% of volume of SO_2 in air with a useful analytical range of 0.01–6%. The quenching efficiency of SO_2 was about 26 times higher than that of oxygen. The response time of the sensor towards gases was 3–60 s for 90% of the total signal change. Gases like nitrogen, carbon monoxide, carbon dioxide and methane were found not to interfere. However, olefins and alkyl or aryl bromides and iodides are potential interferents. Again, a two-sensor technique would be needed to get rid of the oxygen interference.

6 Halide Sensors

Quinolinium and acridinium compounds are dynamically quenched by halide and pseudohalide ions [9]. Therefore, these indicators have been immobilized covalently onto a glass surface. They are typically excited at 350–360 nm and their emission can be measured at 450–500 nm. The detection limits of an acridinium ion-based sensor were 0.15 mM for iodide, 0.40 mM for bromide and 10.0 mM for chloride. Figure 7.5 shows the relative fluorescence intensities of sensors based on both indicator types as a function of the iodide, bromide and chloride concentration. Sulfite, isothiocyanate, cyanide and cyanate also quench the fluorescence of the indicators, whereas sulfate, phosphate, perchlorate and nitrate do not. The average response time for achieving 95% of the total signal change was around 40 s for halide solutions pumped through a flow-through cell.

7 Humidity Sensor

It was found that the fluorescence of certain silicagel-adsorbed perylene dyes is quenched by water vapour [10]. By using perylenetetracarboxylic acid N-alkyl and N-arylimides, a total signal change of 95% was observed in going from 0 to 100% relative humidity. These dyes can be excited at about 490 nm and show emission maxima at approximately 545 nm. The response times depended on the relative humidity of the sample gas and varied from 5 to 10 min. However, the response functions were not linear and molecular oxygen was a strong interferent. In addition, the recovery of the sensor was slow once it had been exposed to 100% relative humidity.

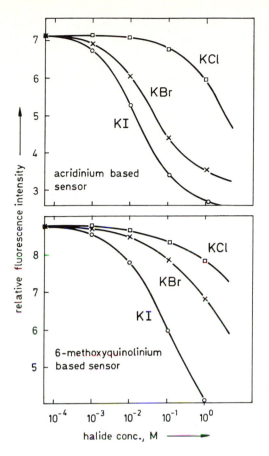

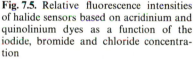

Fig. 7.5. Relative fluorescence intensities of halide sensors based on acridinium and quinolinium dyes as a function of the iodide, bromide and chloride concentration

8 Oxygen Sensors as Transducers

In chemical or biochemical reactions, where molecular oxygen is consumed or produced, an oxygen sensor may be used as a transduction element to get an optically detectable signal. Here, the consumption or production of oxygen results in a change in fluorescence intensity which may be related to the concentration of one of the substrates of the reaction. A catalyst, which catalyzes the desired reaction, can be immobilized close to the oxygen-sensitive dye or onto the surface of the oxygen sensor. If the catalyst is a biologically active material, a biosensor is obtained. The most frequently used biocatalysts are enzymes, because they provide high specificity and high reaction rates.

8.1 Hydrogen Peroxide Sensors

Hydrogen peroxide shows only poor quenching properties. For example, it is a dynamic quencher of the fluorescence of ruthenium(II)-tris(bipyridyl), but its quenching efficiency is much lower than that of oxygen (11). However, hydrogen peroxide can be decomposed by using the enzyme catalase or by using fine silver powder as the catalyst (Eq. 3). The amount of oxygen produced can be measured and related to the hydrogen peroxide concentration.

$$2H_2O_2 \xrightarrow{\text{catalase or silver}} 2H_2O + O_2. \tag{3}$$

Catalase was co-immobilized with the indicator dye on silica gel, which was embedded in a silicone matrix, whereas silver was applied in a thin silicone film on the surface of the oxygen-sensitive layer [11]. The sensors showed an analytical range of 0.1–10 mM of hydrogen peroxide and response times of 2.5–5 min. The silver-based sensor exhibited a better long-term stability than the catalase-based sensor, especially when the sensors were stored dry.

8.2 Glucose Sensors

Glucose is colourless and does not display an intrinsic fluorescence in the visible region of the spectrum. Therefore, it is not easy to detect glucose with optical methods directly. The enzyme glucose oxidase catalyzes the oxidation of glucose by molecular oxygen as the second substrate. Hydrogen peroxide and gluconolactone are formed (Eq. 4). The consumption of oxygen can be made visible again with the help of an oxygen sensor. The signal changes obtained can be related to the glucose concentration of the sample.

$$\text{D-glucose} + O_2 \xrightarrow{\text{glucose oxidase}} \text{D-gluconolactone} + H_2O_2. \tag{4}$$

For example, glucose oxidase was immobilized covalently onto a nylon membrane which was fixed on the oxygen-sensitive layer consisting of solubilized decacyclene dissolved in silicone [12]. This type of glucose biosensor showed an analytical range of typically 0.05–3 mM glucose with response times of 1–3 min for 90% of the total signal change. The system was improved by replacing the rather thick nylon membrane by a glutaraldehyde-crosslinked enzyme layer with an actual thickness of a few micrometers [13]. The glucose sensors obtained with this method had response times of 20–40 s only for 90% of the total signal change, and analytical ranges of approximately 0.02–0.8 mM of glucose (Fig. 7.6).

8.3 Other Biosensors

In principle, the sensing scheme presented for glucose is applicable to all enzymes which consume molecular oxygen during their enzymatic reaction. These are mainly oxidases and oxygenases. A number of such enzymes have been tested till now and a few have shown to be useful. The main problem arises

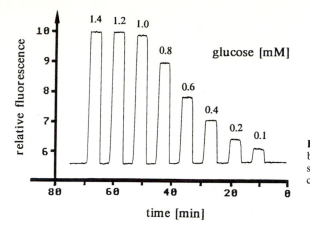

Fig. 7.6. Response of an enzyme-based glucose biosensor towards samples with various glucose concentrations

Table 7.1. Optical biosensors based on oxygen optodes as the transducers

Substrate	Enzyme	References
D-glucose	D-glucose oxidase	[12, 13, 14]
L-lactate	L-lactate monooxygenase	[15]
L-lactate	L-lactate oxidase	[16]
cholesterol	cholesterol oxidase	[17]
ascorbic acid	ascorbic acid oxidase	[18]
L-glutamate	L-glutamate oxidase	[19]
hypoxanthine	xanthine oxidase	[20]
ethanol	alcohol oxidase	[20]
tyrosine	tyrosine oxidase	[20]
phenol	phenol oxidase	[20]
catechol	catechol monooxygenase	[20]
urea	uricase	[20]

with the long-term stability of the enzyme. Table 7.1 shows a list of optical biosensors based on oxygen optodes as the transducers which have proved to be working.

9 Applications

Biosensors can be useful for determination of metabolites in biomedical analysis such as glucose, lactate and cholesterol, in blood, serum and urine. Oxygen has to be measured in blood and breath gas analysis, whereas the detection of the inhalation narcotic halothane is of importance in anesthesia.

Biosensors have shown to be useful in combination with a flow-injection system in food analysis. Glucose and ascorbic acid have been determined in wine and fruit juice [18, 21], lactic acid in milk products [16] and glutamate, for example in flavour enhancers [19] (Fig. 7.7). With the help of flow injection

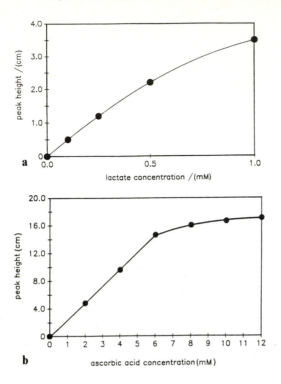

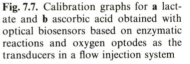

Fig. 7.7. Calibration graphs for **a** lactate and **b** ascorbic acid obtained with optical biosensors based on enzymatic reactions and oxygen optodes as the transducers in a flow injection system

analysis it is possible to overcome a great number of interferences, because the sample is injected into a buffer carrier stream of defined pH, ionic strength and composition. Furthermore, problems with a limited analytical range of the sensor can be overcome by varying the amount of injected sample or the dilution factor of the system [21].

In biotechnology, optical sensors for oxygen and biosensors with an oxygen optode as the transducer may be useful for on-line monitoring of a fermentation process. Some typical substrates and products of fermentations (besides oxygen) are glucose, lactate and ethanol.

A field of major future interest is environmental monitoring. Sulphur dioxide has to be monitored in air and in combustion gases (e.g. in power plants). Halide ions, alkyl and aryl halides have to be determined in ground and waste water. Measuring oxygen provides information about the biological conditions of rivers, lakes and oceans.

10 Future Aspects

Unfortunately, no suitable sensor is available for a large number of analytes of interest at present. But there exists a big variety of potential indicator candidates which could be used in optical sensors. Table 7.2 gives an overview on various

Table 7.2. Fluorophores and their quenchers

Analyte (quencher)	Fluorophor	References
heavy metal ions $(Cu^{2+}, Ag^+, Pb^{2+}, Hg^{2+})$	6-methoxyquinolinium and acridinium ions	[22]
lanthanide(III) ions	calcein blue	[2]
iodine, bromine	rubrene, α-naphthoflavone	[23]
cationic detergents (quarternized ammonium, thiazolium and pyridinium ions)	1-amino- and 1-alkoxy-pyrene-trisulfonates	[2]
aliphatic and aromatic amines	anthracene, naphthalene	[2]
organic sulfides	6-methoxy-4-methyl-2-phenyl-quinoline-4-methanol	[2]
acyclic and cyclic dienes and alkenes	9,10-dicyanoanthracene, pyrene, naphthalene	[2]
nitroaromatics	rhodamins, porphyrins	[2]
thiamine	1-amino-pyrene-trisulfonate	[2]
pyridine	skatole	[24]
nucleosides, nucleotides (AMP, GMP, TMP, CMP)	acridinium ions, proflavine, acridine orange, 9-aminoacridone	[2]

fluorophores and their quenchers. One of the main problems is to find indicators with both, a high sensitivity and a high selectivity for a certain analyte at the same time.

11 References

1. Lakowicz JR (1983) Principles of fluorescence spectroscopy. Plenum, New York
2. Wolfbeis OS (ed) Fiber optic chemical sensors and biosensors, CRC Press, Boca Raton
3. Wolfbeis OS, Posch HE, Kroneis HW (1985) Anal Chem 57: 2556
4. Schaffar BPH, Wolfbeis OS (1989) Proc SPIE 990: 122
5. Wolfbeis OS, Leiner MJO, Posch HE (1986) Mikrochim Acta III: 359
6. Lippitsch ME, Pusterhofer J, Leiner MJP, Wolfbeis OS (1988) Anal Chim Acta 205: 1
7. Sharma A, Wolfbeis OS (1987) Spectrochim Acta 43A: 1417
8. Wolfbeis OS, Sharma A (1988) Anal Chim Acta 208: 53
9. Urbano E, Offenbacher H, Wolfbeis OS (1984) Anal Chem 56: 427
10. Posch HE, Wolfbeis OS (1988) Sensors and Actuators 15: 77
11. Posch HE, Wolfbeis OS (1989) Mikrochim Acta I: 41
12. Trettnak W, Leiner MJP, Wolfbeis OS (1988) Analyst 113: 1519
13. Schaffar BPH, Wolfbeis OS (1990) Biosensors & Bioelectronics 5: 137
14. Moreno–Bondi MC, Wolfbeis OS, Leiner MJP, Schaffar BPH (1990) Anal Chem 62: 2377
15. Trettnak W, Wolfbeis OS (1989) Anal Lett 22: 2191
16. Dremel BAA, Trott–Kriegeskorte G, Schaffar BPH, Schmid RD (1989) In: Schmid RD, Scheller F (eds) Biosensors, VCH Verlagsgesellschaft, Weinheim, p 225 (GBF Monographs, vol 13)
17. Trettnak W, Wolfbeis OS (1990) Anal Biochem 184: 124
18. Schaffar BPH, Dremel BAA, Schmid RD (1989) In: Schmid RD, Scheller F (eds) Biosensors VCH Verlagsgesellschat, Weinheim, p 229 (GBF Monographs, vol 13)
19. Dremel BAA, Schmid RD, Wolfbeis OS (1991) Anal Chim Acta 248: 351
20. Schaffar BPH (1988) (unpublished results)
21. Dremel BAA, Schaffar BPH, Schmid RD (1989) Anal Chim Acta 225: 293
22. Wolfbeis OS, Trettnak W (1987) Spectrochim Acta 43A: 405
23. Wolfbeis OS (1988) In: Schulman SG (ed) Molecular luminescence spectroscopy, methods and applications (part 2). John Wiley, New York, p 235
24. Sharma A, Wolfbeis OS, Machwe MK (1989) Anal Chim Acta 230: 213

Part 2. New Applications of Fluorometry

8. Fluorescence in Forest Decline Studies

H. Schneckenburger[1,2] and W. Schmidt[1,3]

[1] Fachhochschule Aalen, Fachbereich Optolektronik, Beethovenstr. 1, D-7080 Aalen, Germany
[2] Institut für Lasertechnologien in der Medizin an der Universität Ulm, Postfach 4066, D-7900 Ulm, Germany
[3] Universität Konstanz, Fakultät für Biologie, D-7750 Konstanz, Germany

1 Introduction

Plants have to cope with various stress factors such as environmental pollutants, mineral deficiencies, climate or parasite infection giving rise to the current phenomenon of "forest decline". To assess the various impacts of these stress factors, complex biochemical procedures are mostly utilized demanding damage to the specimen and fairly high sample concentrations. In contrast to these, spectroscopic methods offer many advantages. They are fast, non-invasive, easy to perform, highly selective for specific plant pigments, and require only small samples.

So far measurements of chlorophyll fluorescence in the research of forest decline were limited to the detection of emission spectra and kinetics [1] under continuous illumination. The so-called *Kautsky*-measurements are based on the fact that chlorophyll fluorescence of dark-adapted organisms increases to a maximum value f_{max} at the beginning of light exposure and then – during the onset of photosynthesis – decreases within about 5 min to a steady state value f_s. The ratio $(f_{max} - f_s)/f_s = f_d/f_s$ may therefore reflect the photosynthetic activity. This ratio, however, is only a "global" measure including all steps of photosynthesis.

Therefore, *picosecond spectroscopy* was adopted to select the primary steps of energy transfer within the two photosystems [2–6]. After excitation by picosecond laser pulses the fluorescence of the antenna chlorophyll molecules is expected to have a longer decay-time, if the energy transfer to the reaction centers is obstructed. A longer-living fluorescent component as described previously [7] was therefore evaluated for different kinds of spruce needles over a period of about 2 years. Picosecond experiments – combined with microspectrofluorometric studies – were also carried out in the blue and green spectral regions, where the fluorescence emission of specific coenzymes such as NAD(P)H and flavins has been reported [8, 9].

The unexpected phenomenon of *delayed light emission* of photosynthetic organisms dates from the year 1951, when it was discovered by Strehler and Arnold [10]. As compared to prompt fluorescence, delayed light emission turns out to be more complex and more susceptible to change in experimental and particularly plant-physiological conditions. Delayed luminescence is indispensibly dependent on a functioning thylakoid membrane including electrical (ϕ)

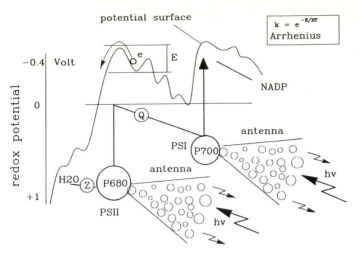

Fig. 8.1. Origin of the various prompt and delayed luminescence signals investigated are the photosynthetic pigment systems "PSI" and "PSII". Kinetics of delayed luminescence are intimately correlated with "potential surfaces" as sketched and determined by Arrhenius' equation

and pH-gradients and is a property of the photo*system*(s) rather than of a single molecular entity (Fig. 8.1). Mainly based on spectral evidence, the majority of researchers attribute delayed luminescence *exclusively* to photosystem II (PS II), i.e. a reverse electron flow:

Absorption: $ZChlQ$ (open state, O) $+ \varepsilon_F \rightarrow Z^+ChlQ^-$ (closed state, P),

Luminescence: Z^+ChlQ^- (closed state, P) $\rightarrow ZChlQ$ (open state, O) $+ \varepsilon_F$

with Z = electron donor, Chl = chlorophyll P_{680}, Q = chinoid electron acceptor and ε_F = exciton. Occasionally the "delay" is explained in terms of repetetive "retrapping" of excitons by the now open reaction center of PS II. However, there is a reasonable body of literature in favor of including also PS I leaving this question unsettled so far [11]. In principle, delayed luminescence reflects the same phenomenon as the so-called glow curves [12] as related to specific activation energies of various electron steps within the photosystems.

Much of the difficulty and uncertainty encountered in compiling and interpreting luminescence data stems from the wide choice of experimental parameters. Luminescence is observed in a large time domain ranging from microseconds up to minutes. It is observed either by the *single-shot* method as induced by a short light pulse, or by a repetetive mode, i.e. by sequential induction and measurement utilizing a phosphoroscopical set-up ("induction kinetics"). Both experimental approaches were used in the present studies.

2 Materials and Methods

2.1 Plant Materials

The results reported were obtained in a joined project under the auspices of PEF (cf. acknowledgement) with spruces at two locations (Fig. 8.2). (1) Measurements at *Edelmannshof* (Welzheimer Wald) are part of a so-called "exclusion experiment" testing for the impact of air pollutants such as O_3, NO_2 and SO_2 in so-called *Open-Top-Chambers* (OTCs). Two spruces (labeled "A" and "B") we analyzed were exposed to filtered air (excluding these pollutants), two other spruces, growing within similar OTCs, but being exposed to non-filtered air (labeled "C" and "D") served as a control. (2) Measurements at *Freudenstadt* (black forest) were performed on 12 individual spruces belonging to different damage classes and accessible from high wooden structures (*Picea abies.* approx. 15 m high and 40–50 years old, age classes 89/90 and 90/91, cf. Table 8.1). For the purpose of standardization, only samples from the 7th whorl from top were taken. Since we are aiming at a convenient procedure of *early* detection of damage in conifers, all measurements were carried out with needles plucked from small, *perfectly green and healthy* looking twigs without any *visible* damage (of course, yellowish needles as mostly found on spruces of damage class III will give quite different results).

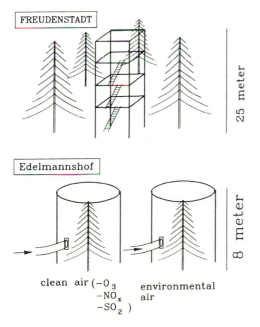

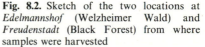

Fig. 8.2. Sketch of the two locations at *Edelmannshof* (Welzheimer Wald) and *Freudenstadt* (Black Forest) from where samples were harvested

Table 8.1. Bonitation in terms of needle loss and yellowing (Seemann, personal communication) and chlorophyll content of spruces (2nd age class) at the location Freudenstadt in Oct. 1990

Tree No.	Needle loss (%)	Yellowing (%)	Chl-content µg/g fw	Damage class
1	40	55	640	3
2	45	50	1220	3
6	50	40	650	3
7	15	5	970	1
12	15	10	980	1
31	15	20	1100	1
13	0	0	1650	0
14	5	5	940	0
15	5	0	1300	0
16	5	10	1310	0
17	10	5	1120	0
24	10	5	1050	0

2.2 Measuring Devices

Various set-ups for the detection of prompt and delayed fluorescence have been recently developed. Stationary fluorescence measurements of the individual spruce needles (Kautsky measurements) were carried out using a fiber-optic sensor, including an air-cooled argon ion laser (operated at 458 nm) and an avalanche photodetector [13].

Picosecond fluorescence decay kinetics were measured by means of a red (668 nm) or a frequency-doubled UV laser diode (390 nm) with subsequent single photon analysis (Fig. 8.3, for details see [7]). The pulse energy of the laser-diode ranged between 10^{-14} J (390 nm) and 10^{-11} J (668 nm), a repetition rate between 100 kHz and 1 MHz was adjusted. The excitation wavelengths were selected such that they were close to an absorption maximum of chlorophyll molecules (668 nm) or some blue or green fluorescent pigments (390 nm). The fluorescence was detected in the spectral ranges of 450 ± 20 nm, 500 ± 20 nm and 695–800 nm. Additional measurements with narrow-band detection of the individual chlorophyll bands are described in the previous article [7]. The photon signals as detected by the photomultiplier were amplified and time-correlated with the trigger pulses of the laser diode using two constant fraction discriminators and a time-amplitude converter. All photon events were summarized in a multichannel analyzer incorporated in a personal computer. The fluorescence decay curves thus obtained were deconvoluted using the apparative response function and fitted by a superposition of several exponential components. No more than 3 components were simultaneously detected. The time resolution after deconvolution was below 100 ps.

Two different set-ups were utilized for the detection of delayed luminescence in the two measuring modes. In one mode induction and decrease of delayed

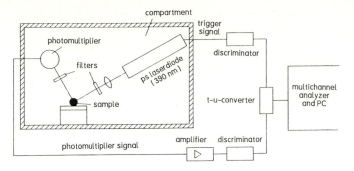

Fig. 8.3. Experimental setup for time-resolved fluorescence spectroscopy based on a frequency-doubled picosecond laser diode and a single photon counting device

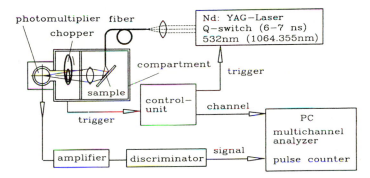

Fig. 8.4. Experimental system for measuring the induction and decrease of delayed luminescence of spruce needles during quasi-continuous irradiation by a Nd:YAG laser (532 nm)

luminescence were measured using a quasi-continuous irradiation by short pulses of a frequency-doubled Nd:YAG laser (532 ns, 6–7 ns, repetition rate 10 Hz). The luminescence decay was detected within a time window of 10–90 ms after each laser pulse by using an appropriate chopper disk (Fig. 4). Immediately before the measurement started, some 25 individual needles from the various spruces were plucked, closely packed side by side on a black aluminum sample holder (glass cannot be used since it shows high intrinsic delayed luminescence) using a double-sided sticky tape, presenting a "green carpet" of always approximately 2 cm^2 with 30° inclination towards laser-light. A Hall sensor fixed on this disk triggered both the laser and the time-channel of signal detection (*multichannel scaling*). After 20 min of dark adaptation the luminescence signal was observed during a period of 60 s. 1000 laser pulses with a light energy of 800 μJ/pulse were applied during this time. The decrease of *long-term delayed luminescence (LDL)* of light-adapted spruce needles in the range up to 20 s following a single light pulse was measured with the apparatus described in detail previously [14]. In essence the sample is irradiated for a few seconds, and immediately afterwards the luminescence decay is monitored. The whole set-up and analysis is largely automatized and computer-controlled.

Parallel to all luminescence measurements the chlorophyll concentrations of needles from the same twigs were determined by absorption spectroscopy following pigment extraction in acetone [15], Table 8.1.

Additional fluorimetric measurements of microscopic needle areas were performed with the microspectrophotometer UMSP 80 described elsewhere [16].

3 Results and Discussion

3.1 Chlorophyll Content

Very low chlorophyll concentrations were measured for the most declined spruces no. 1 and 6, whereas a comparatively high value was obtained for the damaged spruce no. 2. The values of the less damaged spruces no. 7, 12 and 31 were only slightly smaller than those of the "intact" trees. The chlorophyll concentrations of all spruces usually showed seasonal variations within $\pm$ 150 µg/g fresh weight – higher deviations (values fluctuating between 600 and 1100 µg/g fresh weight) were obtained for the spruces no. 17 and 31. A unique continuous increase of chlorophyll content (from 1100 in Nov. 1989 to 1700 µg/g fresh weight in April 1991) was measured for spruce no. 13 which, in contrast to all other trees, showed pronounced dark green needles in spring 1991. The chlorophyll concentrations of the needles (ac 89/90) of all spruces at Edelmannshof were in the range of 1600–2000 µg/g fresh weight; only spruce C showed maximum values of 2200 µg/g fresh weight in summer and autumn 1990.

3.2 Fluorescence Spectra

Fluorescence spectra were recorded from whole needles as well as from microscopic spots of the needle surface. Microscopic measurements showed a pronounced blue-green fluorescence from the stomata – therefore object fields from the stomata and from outside the stomata of 16 µm in diameter were selected. Based on the extensive investigations of Zeiger [17], this blue-light induced fluorescence is most likely correlated with a physiological blue-light photoreceptor in higher plants with no relation to photosynthesis.

Figure 8.5 shows the microscopic spectra at excitation wavelengths of 365 $\pm$ 5 nm (a) and 435 $\pm$ 5 nm (b). In the first case a broad emission band at 440–475 nm and a shoulder around 500 nm were detected. Similar spectral bands were recently found for yeast cells [18] and tentatively attributed to reduced NADH (nicotinamide-adenine-dinucleotide) and (oxidized) flavin

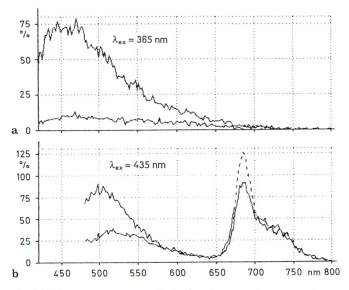

Fig. 8.5. Fluorescence spectra of microscopic areas of spruce needles at excitation wavelengths of **a** 365 ± 5 nm and **b** 435 ± 5 nm. *Full lines*: detection from the stomata; *broken lines*: detection from outside the stomata; spectral resolution 10 nm

molecules, respectively. At 435 nm excitation the green emission band at 500–520 nm and the characteristic chlorophyll fluorescence at 660–750 nm were detected. The chlorophyll fluorescence was higher when measured from outside the stomata, whereas the blue and green spectral bands were more pronounced when detected from inside the stomata. In addition, the green emission band was slightly blue-shifted within the stomata (maximum around 500 nm) as compared with outside the stomata (about 520 nm). The spectra shown in Fig. 8.5 were typical for all trees at the location of Edelmannshof and Freudenstadt; however, the intensity of the chlorophyll band varied and reflected the different chlorophyll concentrations of the individual trees.

3.3 Picosecond Kinetics

Picosecond fluorescence decay kinetics could only be measured from whole needles using an excitation wavelength of 390 nm (Fig. 8.3) or 668 nm . Nevertheless, at 390 nm excitation a large part of the fluorescence detected in the blue and green spectral ranges might arise from the stomata and therefore depend on environmental factors. Decay curves were measured in the wavelength ranges of 450 ± 20 nm and 500 ± 20 nm. They were bi-exponential in the first and fairly tri-exponential in the second case with the time constants listed in Table 8.2. In the blue spectral range only the longer-lived component $T_2 = 2$ ns was fairly in agreement with values obtained from NADH solutions; in the green range the

Table 8.2. Fluorescence decay times after picosecond pulse excitation at 390 nm

Compound No.	Decay time at 450 ± 20 nm	Decay time at 500 ± 20 nm
T_1	$80 - 100$ ps	$100 - 300$ ps
T_2	2.0 ± 0.5 ns	1.5 ± 0.5 ns
T_3	—	5.5 ± 1.0 ns

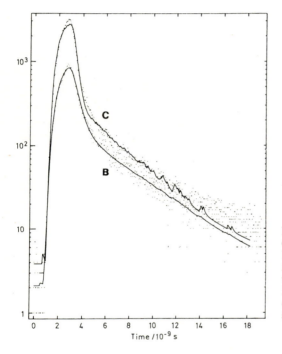

Fig. 8.6. Fluorescence decay curves of 6–8 individual needles from the spruces B (*lower curve*) and C (*upper curve*) at *Edelmannshof* in Jan. 1991. Excitation wavelength 390 nm, detection range 500 ± 20 nm. The corresponding curve of spruce A was similar to B, the curve of spruce D similar to C

time constant T_3 agreed with the decay time of riboflavin ($T = 5.65$ ns [9]) and flavin mononucletide (FMN, $T = 5.2$ ns) in aqueous solution (considerably shorter lifetimes are reported in literature for flavoproteins [9]). It appears that the relative intensities of the individual components depend on the environmental conditions of each tree. This is demonstrated in Fig. 8.6, according to which the shorter-lived components of the spruces C and D (environmental air) are more intensive than those of the spruces A and B (filtered air). These differences, however, were less pronounced in the subsequent experiments with needles from Edelmannshof and Freudenstadt.

Picosecond kinetics of *chlorophyll fluorescence* have been reported recently [7] using an excitation wavelength of 668 nm and a detection range of 695–800 nm. Three components could be resolved with time constants of

$T_1 = 100–200$ ps, $T_2 = 300–500$ ps and $T_3 = 2.0–3.5$ ns . In agreement with literature [2, 6] the latter was attributed to chlorophyll molecules surrounding "closed" reaction centers of photosystem II. The relative intensity I_3 of this component is plotted in Fig. 8.7 for the spruces in Freudenstadt as obtained at different time periods. It turned out that in general I_3 was higher during the summer than during the winter period, and that I_3 reflected the state of damage rather well. Even green needles of the slightly damaged spruces 7, 12 and 31 showed increased values of I_3 as compared with the undamaged spruces (continuously up to April 1991; a decrease of I_3 for the spruces 7 and 12 in summer 1991 remains to be investigated further).

The seasonal course of I_3 was studied more in detail for the spruces at Edelmannshof (Fig. 8.8). Maximum values of this intensity were found in spring and summer. In March–June 1990 these values were higher for the spruces A and D than for B and C, which indicates a possible defect of A and D. Only in autumn 1990 parasitic infection of spruce A by *Physokermes hemicryphus* and drought stress of spruce D became obvious. A further parameter for an early damage was the ratio f_d/f_s between fluorescence decrease and steady state fluorescence of chlorophyll under continuous illumination (*Kautsky curves*). This ratio showed an annual course as has been reported in literature [19, 20].

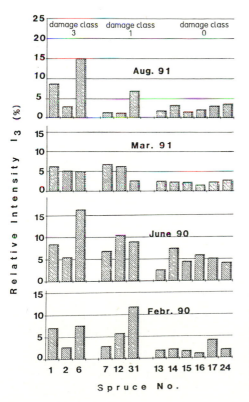

Fig. 8.7. Relative intensity of the long-lived component to chlorophyll fluorescence obtained in each case from 5–10 green needles of the second age class of 12 spruces in Freudenstadt between Febr. 1990 and Aug. 1991. Reproducibility of individual values within ± 10%

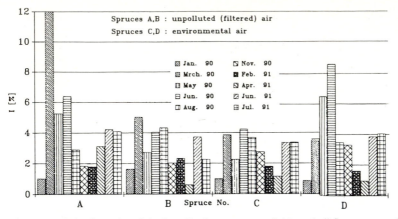

Fig. 8.8. Relative intensity of the long-lived component of chlorophyll fluorescence obtained from green needles of the spruces A, B, C and D at *Edelmannshof* (Jan. 90–Apr. 91: age class 1989; June–Sept. 91: age class 1990). Reproducibility of individual values within ± 10% (the value of spruce A in March 1990 exceeds the vertical axis)

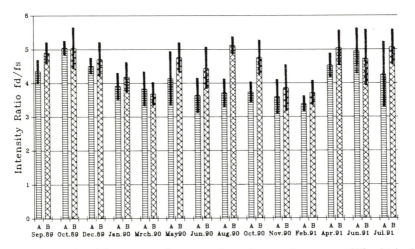

Fig. 8.9. Ratio of fluorescence decrease and steady state fluorescence fd/fs obtained from the *Kautsky measurements* of the spruces A and B (second age class of needles) at *Edelmannshof* during a period of 2 years. Standard deviations result from 12 individual needles in each case

Deviations from this course were found for the spruces A (Fig. 8.9) and D (data not shown) during May – October 1990. This would again imply that damage to the photosynthetic system could be detected in a very early stage. A specific difference between the spruces exposed to filtered (A, B) and environmental air (C, D) – as measured by time-resolved spectroscopy in the green spectral range – was not detected for chlorophyll fluorescence.

3.4 Long Term Delayed Luminescence (LDL)

LDL kinetics, as induced by red light – reflecting a relatively large signal to noise ratio – are perfectly fitted by a three-exponential superposition (Fig. 8.10):

$$A_{LDL}(t) = A_f e^{-kft} + A_m e^{-kmt} + A_s e^{-kst},$$

i.e. by contribution of a "fast" $[A_f]$, a "medium" $[A_m]$ and a "slow" $[A_s]$ component with the corresponding decay constants k_f, k_m and k_s. Time ranges of "fast" and "medium" components are chosen manually prior to the analysis in order to optimize the total fit by conventional χ^2-Test. Whereas reaction constants k are *absolute* values, the contributing amplitudes A are presented as *relative* values summing up to 100% at time zero of the kinetics. Standard measuring errors of the mean are typically largest for k_f ($\pm 8\%$, since these components are monitored for less than 1 s), smaller for k_m, A_f and A_m (2–3%), for k_s und A_s as little as 1%. It should be noted that amplitudes and reaction constants are entities completely *independent* of each other, even if derived from the same kinetics.

One remark appears to be cogent: We believe that in general, particularly for unorthodox photobiological kinetics, the analysis in terms of exponential components is a reasonable procedure of classification (similar to Fourier decomposition). However, a clearcut correlation of single components with biophysical (molecular/membraneous) entities does not necessarily exist. Typically, when the mathematical analysis improves, more and more "components" can be resolved!

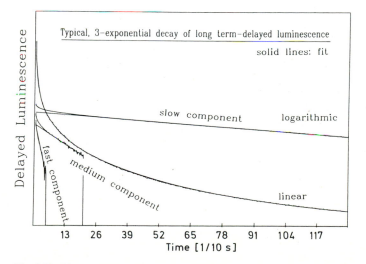

Fig. 8.10. Three-exponential analysis of a typical long term luminescence decay kinetics. The *solid lines* are calculations as based on the individual 6 parameters: three amplitudes (A_f, A_m, A_s) and the corresponding decay constants (k_f, k_m, k_s)

Figure 8.11 shows a complete 16 months time course of the amplitude of the "fast" component of LDL ("A_f") of spruces no. A and B (purified air) and C and D (environmental air) in the OTC-experiments at *Edelmannshof* for both age classes, i.e. the latest one and the previous one. There is a typical, dramatic annual time course of this component demonstrating that individual measurements at specific sampling times are futile. In general, A_f-values are obviously smaller in spruces no. A and B compared to spruces C and D. In addition, younger needles (ac 89/90) always show a smaller contribution of the "fast amplitude" than older ones (ac 90/91).

Another clearcut example (Fig. 8.12) as taken from the series of measurements of the 3 spruces no. 1, 7 and 13 at *Freudenstadt* supports this phenomenon: A_f-values of spruce no. 13 (damage class 0, the most "healthy tree"), spruce no. 7 (damage class 1) and no. 1 (damage class 3) increase with damage class. Unfortunately, this relationship between damage class/environmental pollution and small A_f-values is not necessarily distinctive and might – in addition – also reflect properties of the individual tree. In contrast to amplitudes, the reaction constants, e.g. the k_s-values, appear to be more stable during the annual time-course, showing less variation between trees (Fig. 8.13); however they appear to be somewhat lower in less damaged trees.

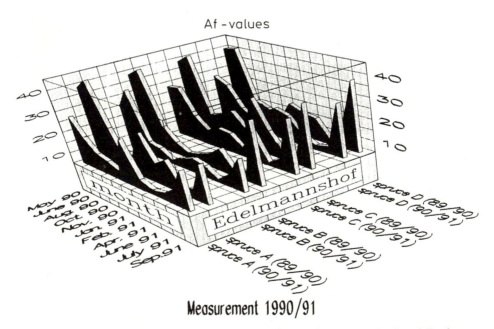

Fig. 8.11. 16-months time course of relative amplitudes A_f of spruces A, B, C and D of two subsequent age classes (89/90 and 90/91) investigated at the location *Edelmannshof*

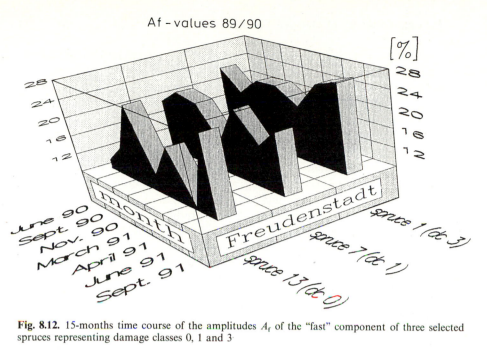

Fig. 8.12. 15-months time course of the amplitudes A_f of the "fast" component of three selected spruces representing damage classes 0, 1 and 3·

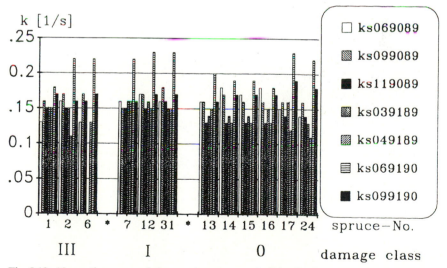

Fig. 8.13. 15-months course of the reaction constant k_s ("slow" component) as obtained from all spruces investigated at the location *Freudenstadt* (age class 89/90)

3.5 Induction Kinetics of Delayed Luminescence

Figure 8.14 shows the *induction kinetics* of delayed luminescence of a light- and a dark-adapted spruce sample. In both cases the kinetics start from close to zero-point of luminescence either to increase to a maximum (using light energies of $> 200 \, \mu J$/pulse) after approximately 10 s and to decline assymptoticly to a steady state (dark-adapted), or to reach the same steady state immediately (light-adapted), independent of the *initial status* of adaptation. The height of the peak maximum of dark-adapted samples depends approximately on the *square root* of pulse energy (light intensity, Fig. 8.15) clearly excluding a (trivial) mono-molecular deexcitation and indicating a *bimolecular* reaction (as for triplet–triplet annihilation in common delayed fluorescence).

The clearcut equivalence of the electrical gradient across thylakoid membranes of isolated chloroplasts and delayed luminescence (as based on salt-jump-induced luminescence) has been demonstrated some 20 years ago by Barber [21]. This offers a simple and fast assay of the development of the photo-induced electrical gradient across the thylakoid membranes which, in turn, immediately reflects the (current) activity of the photosynthetic system. This assumption appears to be justified as tested by induction curves of spruces belonging to different damage classes (Table 8.1, Fig. 8.16). The smaller the damage class, the larger the "induction peak", and the higher the steady state level! It is interesting to note that spruce no. 24 shows kinetics different from those of all other spruces, although this tree did not show any peculiarities otherwise, so far. The more damaged spruces exhibit a smaller specific chlorophyll content [Table 8.1], which, however, *cannot* explain the observed differences.

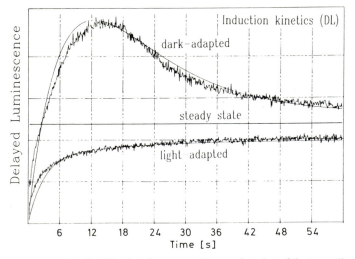

Fig. 8.14. Induction kinetics of spruce needles as a function of the (preceding) level of adaptation, as indicated (solid lines: fits according to the equation given on p. 108)

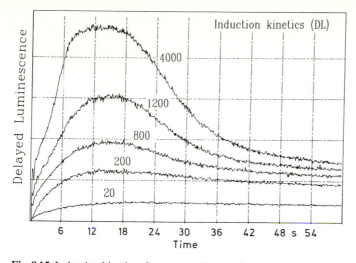

Fig. 8.15 Induction kinetics of spruce needles as a function of energy of the exciting laser pulses (μJ)

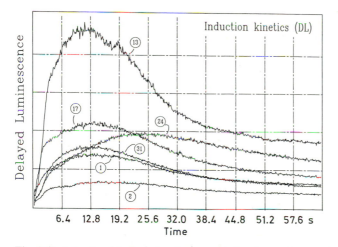

Fig. 8.16. Induction kinetics of various trees belonging to damage class 0 (13, 17, 24), 1 (31) and 3 (1, 2) at the location *Freudenstadt*

The particular time course of induction kinetics cannot be explained since its precise origin is unknown so far. It strongly depends both on the sample itself as well as on the specific experimental set-up utilized. In contrast to the laser-pulse induction used for the present study, Stein et al. [22] employed a measuring cycle of 200 ms composed of 70 ms white DC light induction followed by a 30 ms dark period, 70 ms measurement and another 30 ms dark period. In mathematical terms the general shape of the induction kinetics obtained is relatively simple. It parallels the well-known kinetics of the inter-mediate component "B" in the reaction sequence

$$A \xrightarrow{k_1} B \xrightarrow{k_2} C$$

with the time dependence

$$B = Bo[1 - \exp(-k_1 t)][\exp(-k_2 t) + \text{off}].$$

Only when including an appropriate offset (off) in the second term of the equation above reasonably good fits (Fig. 8.14) with reaction constants ranging between 0.07 to 0.15 s^{-1} (solid lines) are obtained.

4 Resume

Concluding, the various time-resolved luminescence techniques presented (*Kautsky kinetics*, picosecond spectroscopy, long term delayed luminescence and *induction kinetics*) offer the potential of diagnosis of damage of photosynthetic organisms, particularly of trees such as conifers. They probably reveal the statuses of different sites and their (modified) properties. So far nearly exclusively laboratory-bound in vitro, i.e. laborious biochemical wet-procedures de-manding the damage of the specimen and fairly high sample concentrations are utilized in this field. We will now explore further the *origin* of the various luminescence and fluorescence signals in spruces by means of more rigorous wavelengths discrimination for excitation and emission as performed previously with green algae [23], aiming at a clearcut correlation of the type of damage and the resulting kinetic property.

Acknowledgements. This work was supported by "Projekt Europäisches Forschungszentrum für Maßnahmen zur Luftreinhaltung" (PEF). The cooperation of M. Trump is greatfully acknowledged.

5 References

1. Lichtenthaler HK, Buschmann C, Rinderle U, Schmuck G (1986) Radiat Environ Biophys 25: 297
2. Holzwarth AR (1988) In: Lichtenthaler HK (ed) Applications of chlorophyll fluorescence. Kluwer Acad Publ, Dordrecht, p 21
3. Hodges M, Moya I (1988) Biochim Biophys Acta 935: 41
4. Sparrow, Evans EH, Brown RG, Shaw D (1989) J Photochem Photobiol B3: 65
5. Sparrow R, Brown RG, Evans EH, Shaw D (1990) J Photochem Photobiol B5: 445
6. Krause GH, Weis E (1991) Ann Rev Plant Physiol Plant Mol Biol 42: 313
7. Schneckenburger H, Schmidt W (1992) Radiat Environ Biophys 31:73
8. Salmon J-M, Kohen E, Viallet P, Hirschberg JG, Wouters AW, Kohen C, Thorell B (1982) Photochem Photobiol 36: 585
9. Galland P, Senger H (1988) J Photochem Photobiol B1: 227
10. Strehler BL, Arnold W (1951) J Gen Physiol 34: 809
11. Lavorel J (1975) In: Govindjee (ed) Bioenergetics of photosynthesis. Academic Press, New York, p 223

12. Litvin FF, Sineshchekov VA (1975). In: Govindjee (ed) Bioenergetics of photosynthesis. Academic Press, New York, p 620
13. Schneckenburger H, Schmidt W, Hammer P, Hudelmaier K, Pfeifer R (1990) In: Waidelich W (ed) Optoelectronics in Engineering. Springer, Berlin Heidelberg New York, p 403
14. Schmidt W, Schneckenburger H (1992) Radiat Environ Biophys 31:63
15. Lichtenthaler HK, Wellburn AR (1983) Biochem Soc Transact 603: 591
16. Schneckenburger H, Lang M, Köllner T, Rück A, Herzog M, Hörauf H, Steiner R (1989) Lasers Life Sci 4: 159
17. Zeiger E (1984) In: Senger H (ed) Blue Light effects in Biological Systems, Springer Berlin Heidelberg New York, p 484
18. Schneckenburger H, Rück A, Haferkamp O (1989) Anal Chim Acta 227:227
19. Bolhar-Nordenkampf HR, Lechner E (1988) In: Applications of Chlorophyll Fluorescence. Lichtenthaler HK (ed), Kluwer Acad Publ, Dordrecht (NL), p 173
20. Lichtenthaler HK, Rinderle U, Haitz M (1989) Ann Sci For 46, suppl: 483s
21. Barber J (1972) Biochim Biophys Acta 275: 105
22. Stein U, Buschmann C, Blaich R, Lichtenthaler HK (1990) Radiat Environ Biophys 29: 119
23. Schmidt W, Senger H (1987) Biochim Biophys Acta 890: 15

9. Fluorescence Microscopy Studies of Structure Formation in Surfactant Monolayers

Hans Riegler* and Helmuth Möhwald

Institut für Physikalische Chemie, Welder-Weg 11, Universität Mainz, D-6500 Mainz, Germany

1 Introduction

Monolayers of water-insoluble surfactants (Langmuir monolayers) are important model system in basic research for studying the self-organization of organic molecules into two-dimensional layers [1, 2]. These floating films are also the basis for the build-up of complex multilayered structures on solid substrates, so-called Langmuir–Blodgett films (LB-films), which are of considerable interest in fundamental science and promise some future practical applications [3].

Up to about ten years ago the methods of investigation of floating mono-layers were restricted to measurements on a macroscopic scale. The interpretation of "classical" pressure/area isotherms, rheological and surface potential studies yielded only (important) indirect information on the structure and dynamics of these films. A breakthrough in the investigation of floating monolayers, which produced direct information on a micrometer scale occurred around 1983 with the application of fluorescence microscopy [4–6]. Since then, it has been possible to directly visualize molecular aggregations and measure the statics and dynamics of molecular ordering in the floating films. In the meantime the fluorescence microscopic investigation of Langmuir monolayers has become a standard method in many laboratories and has been used to investigate the molecular packing, chain orientation, and domain morphology of many different surfactants in various phases [7–21]. Recently it has also been used to study the adsorption from the aqueous subphase at surfactant monolayers [22, 23]. Fluorescence microscopy is a comparatively simple and inexpensive method and, although recent progress in imaging ellipsometry [24, 25] may eventually lead to its replacement in some area, it will remain an important monolayer investigation technique.

In the following we will shortly outline the measurement principle and then present some microfluorescence studies of monolayers on the plane water surface and during their transfer onto solid substrates. We will restrict ourselves to results obtained for liquid-condensed/liquid-expanded (LC/LE) phase coexistence where this technique has been employed most successfully due to its capability of imaging the LC domains in a LE matrix. This is not a severe restriction because the system conditions of many classical amphiphiles can be adjusted suitably so that they exhibit LC/LE-phase coexistence. Therefore these results are of common value. It will be demonstrated what interactions influence the number, size, shape, and overall arrangement of LC domains. As an extension of the standard microfluorescence technique transfer microscopy has been developed [26] to investigate the impact of the change of monolayer support from water to solid surfaces. Thus, it is shown that the monolayer ordering and composition can be substantially influenced by the LB-transfer. Substrate-mediated condensation is observed and the local monolayer modifications can be described in the framework of classical solidification.

2 Principles of the Microfluorescence Technique

To visualize structures the surfactant monolayer is usually doped with minor amounts (≈ 1 mol%) of fluorescing dye. Figure 9.1 shows a typical pressure/area isotherm with a LC/LE-phase coexistence at pressures above π_c and characterized by a nearly horizontal (high compressibility) region of the isotherm. Typical for the phase coexistence region is the existence of LC domains embedded in an LE-matrix. These domains can be imaged via brightness contrast with a fluorescence microscope due to the different dye solubility in different surfactant phases [4–6]. Usually, the dye solubility in the higher-ordered, more densely packed LC-phase is much lower than in the less ordered LE-phase. Thus, dark LC domains are observed in a bright LE-matrix. If the surfactant itself is fluorescing, the addition of dye is not necessary and the imaging is based on the variation of the fluorescence spectrum and intensity due to local monolayer packing [13, 14]. Fluorescence microscopy yields information on the statics and dynamics of the number, size, shape, and overall local arrangement of LC domains as function of the system parameters (temperature, pressure, etc.).

The transfer microscope (Fig. 9.2) is a special fluorescence microscope which utilizes the concept of the microfluorescence technique to investigate monolayer alterations caused by the Langmuir–Blodgett transfer (LB-transfer) onto solid substrates [26, 27]. The monolayer deposition onto substrates via LB-transfer is accomplished by the sequential (vertical) dipping of the substrate through the monolayer. By means of this process series surfactant monolayers can be

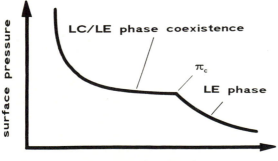

Fig. 9.1. Pressure/area isotherm of a surfactant with a LC/LE phase coexistence regime above π_c

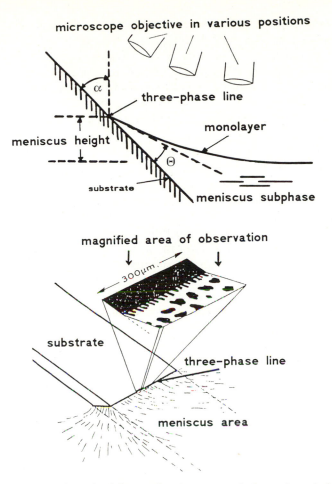

Fig. 9.2. Schematic of the transfer microscope and observation principle

deposited onto the solid substrate as multilayers [28]. For on-line investigations in the three-phase line area the transfer microscope can image any region of the entire monolayer surface including the inclined area in the vicinity of the three-phase line. At the three-phase line the monolayer contacts the solid substrate, where it passes from the water subphase to the solid support. Observations in the three-phase line vicinity yield information on the statics and dynamics of substrate-induced alterations of the monolayer structure.

3 Investigations on the Plane Water Surface

This section presents studies of monolayers on the plane water surface in their LC/LE-coexistence phase. The water surface provides a flat, smooth support for the monolayer where the surfactant molecules are highly mobile and the monolayer can be modified by changing the area per molecule, the ionic strength, etc. Because the trough dimensions are much larger than the domain size, boundary effects can be neglected. The number of domains, their size, shape and overall arrangement is characteristic for the substance and the experimental conditions (temperature, pressure, etc.)

The interplay of the following three types of interactions determines the morphology of the LC/LE-coexistence phase:

1. *Electrostatic interactions.* They are always present because in many cases the headgroups of the surfactants are charged and the charge densities of the LC- and LE-phases are different due to different packing densities. Even if the molecules are uncharged they still have dipole moments resulting in different dipole densities in the LC- and LE-phase. Electrostatic inter-domain interactions are always repulsive and the electrostatic self energy favors the reduction of domain size or the elongation of the domains. Experimentally the electrostatics can be modified by changing subphase ionic conditions.

2. *Line tension.* This energy enforces structures with reduced boundary lengths. Contrary to electrostatics it favors the growth of few larger domains compared to many small ones. It does not contribute to inter-domain interactions but is important for domain nucleation kinetics. Experimentally it can be modified by adding small amounts of surface active impurities (e.g. cholesterol).

3. *Steric effects.* The individual shape of the surfactant molecules can strongly affect the domain shape. Packing restrictions may lead to star- or needle-like domains; molecular chiralities may be reflected in chiral shapes. Experimentally the latter can be studied by comparing racemic mixtures with the pure enantiomer.

3.1 Domain Superlattice

After compression of a monolayer above the pressure π_c small LC domains appear, which grow in size as the pressure is further increased. A domain arrangement typical for many surfactants is shown in Fig. 9.3 [29–31]. At higher domain densities we observe a hexagonal superlattice. The domain size is monodisperse and increases or decreases when the pressure is increased or decreased, respectively. If the surface pressure is changed only slowly, the number of domains remains always constant as long as we remain in the coexistence regime. The number is determined by growth kinetics (cf. Sect. 3.3) which means monolayers as shown in Fig. 9.3 are not in their equilibrium state. Nevertheless, the observed superlattices are usually stable for time periods longer than experimentally accessible (> 24 h). They are stabilized by repulsive forces because inter-domain distances are uniform. This points to electrostatics [37], which is repulsive and tends to reduce the domain size. It has in fact been shown that both together, inter-domain repulsion and minimization of domain size can create hexagonal domain superlattices [31].

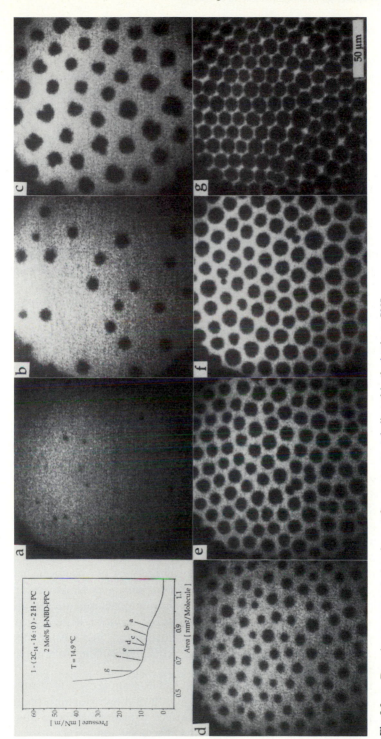

Fig. 9.3a–g. Domain arrangement at various surface pressures as indicated in the isotherm [29]

3.2 Domain Shape and Size

The shape of the domains is determined by the competition between line tension, electrostatic self energy, and molecular packing restrictions. This leads to a great variety of domain morphologies. An example for the temperature dependent interplay of the various energy contributions is shown in Fig. 9.4 [32–34]. The depicted Dimyristoylphosphatidic acid (DMPA) monolayer has been doped with 2 mol% cholesterol which affects the line tension (pure DMPA does not show the temperature behavior described in the following). In the experiment the temperature has been lowered from 15°C to 5.2°C and then again increased to 15°C. At 15°C we observe a hexagonal superlattice of nearly uniformly sized roundish domains. The domain shape is primarily governed by line tension which minimizes the domain boundary, the inter-domain arrangement and domain size is determined by electrostatics. At the end of the temperature-cycle we return to the initial state, hence Fig. 9.4 represents an equilibrium process. Upon cooling to 11.2°C the originally roundish domains begin to form bean-like structures, a process which results in a lamellar structure at 5.2°C. At this temperature we still clearly observe the impact of electrostatics because the structures are equidistant and they do not anneal. However, these anisotropic domains with long boundary lines do not represent a minimum of the line tension any more. Their shape has been explained by noncompensating in-plane dipole moments which cause a temperature dependent anisotropic molecule alignment.

The influence of electrostatic self energy on domain shapes has been studied theoretically [8, 35, 31] and shape transitions between successive modes have been predicted [36]. These size-dependent transitions from round to elliptical and higher mode shapes have in fact been observed [29] and directly investigated for DMPA-domains [31]. For this purpose the effective headgroup charge has been varied via the subphase ion concentration according to Gouy–Chapman–Stern theory. It was observed that the domains are larger and merge with each other at higher Ca^{2+} concentrations. This result agrees qualitatively with the predicted behaviour although the binding of ions affects also the headgroup structure and thus mixes purely electrostatic effects with steric contributions.

Figure 9.5 compares the domain morphology of two isomeric branched chain phosphatidylcholines [29]. These isomers have very similar calorimetric bulk properties and isotherms. However, their domain shapes are very different. One isomer aggregates into crystalline-shaped domains with a nice 6-fold symmetry. The other isomer condensates into round, line-tension optimized domains. This different behavior can be attributed to the different packing properties of the isomers.

The impact of impurities on the kinetics of domain growth is demonstrated by Fig. 9.6 [38, 39]. It shows the chaotic growth of dimyristoylphosphatidylethanolamine (DMPE) domains after the surface pressure has been quickly increased (i.e. a pressure jump above π_c). This affects the domain shapes as follows:

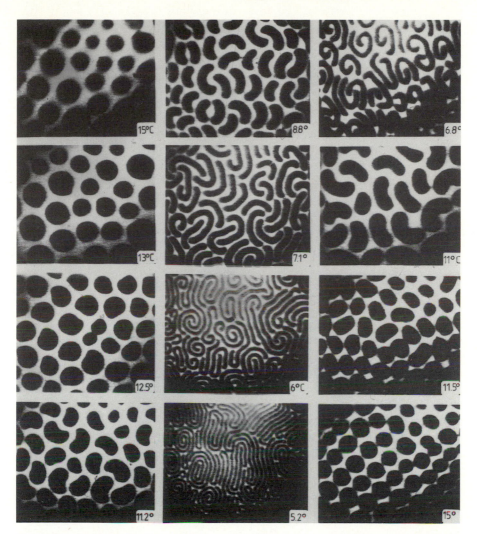

Fig. 9.4. Development of surface textures on cycling the temperature as indicated in the frames. The monolayer contains 2 mol% cholesterol to reduce the line tension [33]

The instantaneous pressure increase induces a diffusion-limited aggregation (DLA). When the monolayer condenses into LC domains, the fluorescent dye is expelled from the LC-phase. The dye transport is limited by diffusion. Thus, if the domain growth proceeds fast, the dye is enriched at the LC/LE boundaries and causes a local increase of π_c due to Raoult's law. This, however, slows down or even stops the condensation which leads to the destabilization of the growth front into fractal dendrites. At constant pressure these dendritic shapes slowly anneal and after typically 30 min, the round equilibrium DMPE-domains are restored [39]. Fluorescence microscopy has been used to determine the

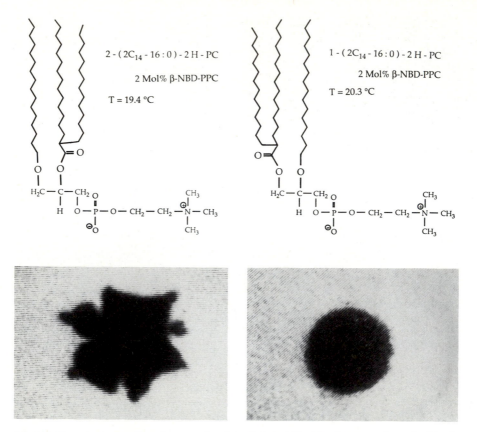

Fig. 9.5. Domain shapes of two isomers of a triple-chain lecithin. The shapes are determined by the different molecular packing properties of the isomers [29]

Haussdorf dimension of these domain to 1.5 in excellent agreement with DLA theory [38].

3.3 Number of Domains

The domain density (number of domains per area) is determined by nucleation kinetics. This has been shown for many different surfactants by varying compression speed, temperature, and impurity concentration. Slow compressions create few, large domains, fast compression generates many, small domains. Once the coexistence pressure π_c has been reached, and a certain domain density has been established, further slow compression induces only the growth of the already existing domains [40, 41]. Fast compressions also induce the growth of the already existing domains but also generate additional small domains. This results in a bimodal size distribution, as shown in Fig. 9.7 [40]. One clearly

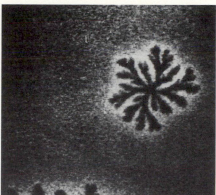

a)   t = 0 sec

b)   t = 1 sec

c)   t = 2 sec

d)   t = 3 sec

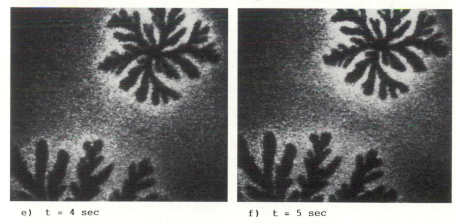

e)   t = 4 sec

f)   t = 5 sec

50 μm

Fig. 9.6a–f. Time development of domain growth following a fast pressure increase. The monolayer contains 0.7 mol% of fluorescence dye causing a diffusion limited domain growth [38]

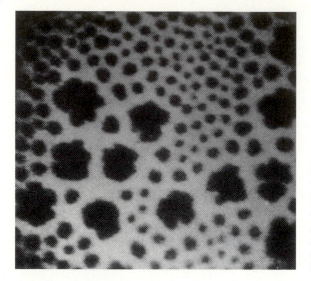

Fig. 9.7. Bimodal size distribution generated by a fast, short pressure increase within the LC/LE phase coexistence regime. The large domains existed before the pressure jump and grew in size after the compression. The small domains were created during the compression [40]

observes two different domain sizes. The large domains were created in the first compression step, the small ones appeared as a consequence of a quick additional pressure jump. The study of the nucleation kinetics as function of temperature and pressure deviation from equilibrium has been used to determine the line tension [13, 40]. The results can be interpreted in terms of classical nucleation theory which is based on the competition between the line tension which hinders the growth of nuclei, and the van der Waals energy which is gained upon molecule aggregation. In the case of monolayers the electrostatic self energy has been treated as an additional term which effectively increases the line tension because it also favors the decrease of aggregates and scales to a first approximation with the linear dimensions of the aggregates [31].

4 Investigations During the Hydrophilic Langmuir-Blodgett Transfer

The next paragraphs address the impact of the transfer of a Langmuir monolayer from the water subphase onto a solid support. We will describe substrate-induced monolayer alterations and phase transitions which were studied with the transfer microscope during the hydrophilic LB-transfer onto amorphous SiO_2 surfaces. The surfactants used in these experiments were DMPE or Dipalmitoylphosphatidylcholine (DPPC), both yielding qualitatively the same results. The investigations were performed by imaging the three-phase line area where the monolayer contacts the solid substrate. By analyzing the domain morphology in the LC/LE-coexistence regime it is shown that the deposition affects monolayer structure and dynamics.

4.1 Substrate-Mediated Condensation with a Stationary Substrate

The substrate surface induces the condensation into the LC-phase. This is revealed by the deposition of a dark, probe depleted monolayer stripe when we compress the monolayer slowly from the LE-phase into the LE/LC-coexistence range (Fig. 9.8) [42, 43]. Before compression the three-phase line (where the monolayer touches the substrate, cf. Fig. 9.2) was allowed to relax into its equilibrium height. During compression the substrate was kept stationary. The growth of the dark monolayer stripe begins already in the LE-phase at a pressure π_c^s below π_c. This substrate-mediated stripe deposition has the following properties:

1. It is not a conventional LB-transfer because we do not move the substrate; only the pressure is increased.
2. The stripe is being deposited while it is growing. This is easily demonstrated by stopping the compression and fully re-expanding the monolayer into the LE-phase. The dark stripe does not melt or re-dissolve into the LE-phase as floating domains do. The three-phase line remains at the position corresponding to the highest pressure applied in the experiment.
3. The stripe spans the entire width of the substrate (typical: 2 cm) and can increase to several 10 µm in width, depending on the maximum applied pressure and some other system parameters (dye concentration, substrate preparation, etc.).
4. The stripe growth and deposition starts at the original (equilibrium) three-phase line and proceeds by lowering the meniscus height as the stripe widens. This lowering of the meniscus height is not simply an equilibration due to the

Fig. 9.8. Stripe deposition created by substrate-mediated condensation. The three-phase line diagonally bisects the area into the deposited monolayer (*left*, dark side including the dark stripe) and the monolayer floating on the meniscus (*right*, bright area including floating domains). The width of the stripe is ≈ 10 µm [42]

increase of the surface pressure. A decrease of the water/monolayer/gas interface energy (compression) results in the meniscus height going up. Our observation of the meniscus height going down can only be explained by a concommittant decrease of the substrate/monolayer/gas interface energy as we reach π_c^s in the course of compression [43, 44].

All these observations indicate that *substrate-mediated condensation* is a consequence of an effectively higher surface pressure on the substrate compared to the water surface. Accordingly, when the surface pressure reaches the threshold value π_c^s the monolayer at the three-phase line starts to condense into the LC-phase concommittant with its deposition. For the systems investigated this threshold pressure π_c^s is lower than the corresponding coexistence pressure π_c on the water surface.

4.2 Morphological Instabilities of the LC Growth Front

If the three-phase line was established below its equilibrium position before compression, the substrate-mediated condensation initiates a destabilizing, floating LC-growth front [45]. The necessary sub-equilibrium meniscus positioning is accomplished by resubmerging the substrate by several 10 µm after the three-phase line relaxation. The already deposited monolayer does not detach from the substrate surface, i.e. the three-phase line is pinned to the substrate surface [46]. Upon compression into the LE/LC-phase coexistence we

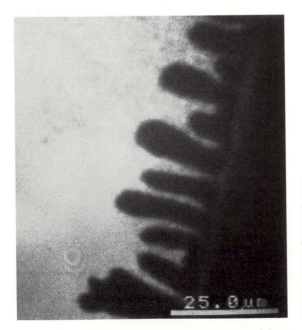

Fig. 9.9. Morphological instability leading to finger-like LC-growth front. The three-phase line diagonally bisects the depicted area into the deposited monolayer region (*right*, dark area) and the floating monolayer area (*left*, bright area including the finger-like domains). The fingers are floating on the meniscus subphase and attached to the substrate at the three-phase line [42]

observe the growth of fingerlike, equidistant domains which are attached to the substrate at the three-phase line and extend floating into the meniscus area (Fig. 9.9). Obviously, this situation differs from the LC-stripe deposition (Sect. 4.1): condensation is restricted to the floating monolayer because deposition is suppressed due to the sub-equilibrium meniscus height. This LC-growth front destabilizes into a finger-like morphology primarily caused by diffusion-limited aggregation (cf. Sect. 3.4) [42]. Crystalline growth anisotropies and electrostatic contributions may also be important. The finger spacing varies with compression speed, dye concentration, substrate preparation, etc.

4.3 Substrate-Mediated Condensation and Local Dye Distribution During Continuous LB-Transfer

The substrate facilitates the condensation of the monolayer into the LC-state as has been shown in the preceding paragraphs. Hence, during a continuous, slow upstroke of the substrate, with the pressure kept constant at some value above the threshold pressure π_c^s for substrate-mediated condensation, we expect a continuous condensation of the monolayer onto the solid surface. As a consequence the monolayer is deposited with a LC/LE ratio which is higher on the solid surface compared to that of the floating film [43, 45].

If the monolayer is doped with dye for fluorescence imaging the substrate-mediated condensation has important consequences on the local monolayer composition as can be seen in Fig. 9.10 [45]. The time sequence ($\Delta t = 2\,\mathrm{s}$) of micrographs shows the monolayer morphology in the vicinity of the three-phase line during a continuous slow substrate upstroke. The depicted situation reflects the dynamic equilibrium, i.e. after some layer has already been deposited. The three-phase line diagonally bisects the micrographs, with the dark areas on the left side showing the deposited monolayer, the right halfs represent the floating monolayer. The deposited monolayer contains no domain-like structures and it is darker than the floating film due to some fluorescence quenching. For the floating monolayer we observe local variations of the LC/LE area-fraction in the vicinity of the three-phase line. Most conspicuous is the absence of any domains in front of the three-phase line, i.e. there is a domain-free gap of about 15 µm width. Further away from the three-phase line we do observe domains, but a careful image analysis reveals that there the LC/LE ratio also varies with distance from the three-phase line. It rapidly increases behind the gap and eventually reaches the equilibrium ratio further out on the water subphase. Because the six micrographs have been taken in 2 s intervals during the continuous transfer we can follow individual domains moving from the upper right to the lower left side. Those domains which come close to the three-phase line melt before they reach the substrate.

These observations can be explained as follows [43]: The substrate induces the deposition of condensed (LC) monolayer. This is a fractional condensation which selectively deposits probe-depleted monolayer because the dye is expelled

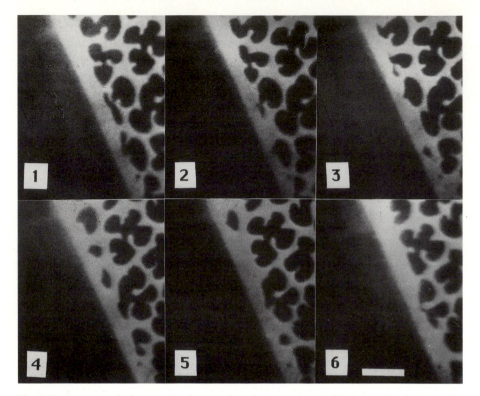

Fig. 9.10. Sequence of micrographs of a monolayer in dynamical equilibrium during its deposition with slow transfer speeds showing the consequence of substrate-mediated fractional LC-deposition (deposition speed = 1 μms^{-1}, time interval between consecutive frames = 2 s). The three-phase line diagonally bisects the monolayer area into the deposited region (dark, left side) and the region floating on the meniscus (bright, right side including the domains). The bar in frame 6 indicates 20 μm [45]

from the LC-phase. As a consequence, the fluorescence dye piles up ahead of the three-phase line. The fractional condensation proceeds continuously during transfer. In the beginning, the dye pile-up may be low, but it will increase and eventually reach a dynamical equilibrium. This is attained when the local dye concentration is so high that the selectivity of the substrate-mediated condensation is just sufficient to condense the monolayer with a dye concentration equivalent to the overall initial inweight of dye. Then the depositing monolayer contains the overall dye concentration and mass conservation is reached. The local (constant) dye pile-up increases the coexistence pressure according to Raoult's law and thus decreases the local LC/LE ratio: If the applied surface pressure lies below the coexistence pressure corresponding to the local dye concentration no domains will exist in this area. This obviously is the case for the domain-free gap ahead of the three-phase line. At the boundary line between the domain-free gap and the area with domains the local dye concentration exactly adjusts the coexistence pressure equal to the applied surface pressure.

Further outside, the dye concentration is not sufficient to melt the domains but the LC/LE ratio is affected according to the increased dye concentration.

The local dye distribution is determined by three processes:

1. The lowering of the coexistence pressure of the substrate (π_c^s) relative to its value on the aqueous phase (π_c), in other words, the selectivity of the substrate surface for the LC-phase.
2. The continuous monolayer (and dye) transport towards the three-phase line with the deposition speed due to the LB-transfer of the monolayer.
3. The diffusional dissolution of the dye towards the floating monolayer caused by the excess dye concentration at the three-phase line.

Process 1 strongly depends on substrate preparation and the monolayer/dye combination used. The "selectivity" determines the amount of dye pile up at the three-phase line. The local functionality of the dye concentration is determined by Process 2. and 3. A model describing a dye concentration profile is shown in Fig. 9.11. The profile depends exponentially on the deposition speed and the diffusion coefficient of the dye in the liquid matrix according to:

$$C(x) = C_\infty + \Delta C_0 * \exp(vx/D)$$

with: $C(x)$ dye concentration at a distance x from the three-phase line, ΔC_0 dye

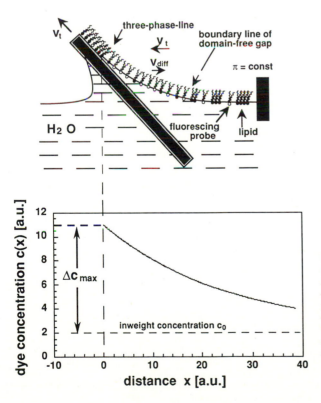

Fig. 9.11. Dye concentration $C(x)$ as function of the distance x from the three-phase line during slow deposition. The plot depicts a model based on a diffusion-regulated dye distribution representing a dynamical equilibrium situation as shown in Fig. 10 [43]

enrichment at the three-phase line, C_∞ dye concentration far away from the three-phase line = original inweight, v deposition speed, x distance to the three-phase line, D diffusion coefficient of the dye in the LE-matrix

The validity of the model has been verified by two ways: Variation of the deposition speed at constant surface pressure and determination of the gap width, and measurement of the gap width as function of the lateral pressure with the transfer speed constant. In both cases the theoretically predicted relations were obtained. With this method the diffusion coefficient of the dye in the fluid matrix has been determined by only measuring the transfer speed and the gap width. The value obtained (13×10^{-8} cm^2/s) agrees well with literature data.

5 Conclusion

Fluorescence microscopy is a formidable technique for investigating the morphology of Langmuir monolayers on a micrometer scale. For this purpose the monolayers are doped with minor amounts of a fluorescence probe and the lateral dye distribution is investigated. This technique has been most successfully applied to monolayers in their LC/LE-phase coexistence where dark probe-depleted LC-domains are observed in a bright, probe enriched LE-matrix. The morphology and lateral arrangement of these domains allows one to draw conclusions on the interactions which determine the molecular ordering in the monolayer. The various contributions of electrostatic interdomain repulsion, electrostatic self energy, line tension, and steric packing restrictions have thus been studied in some detail.

The extension of the microfluorescence technique to monolayer observations in a typical LB-transfer configuration reveals a significant restructuring of the monolayer on changing the monolayer support from the aqueous subphase to the solid support. Substrate-mediated condensation can induce local phase transitions concommittant with the monolayer transfer. The three-phase line, where the monolayer touches the substrate, can also serve as a nucleation line where a floating LC-growth front originates which destabilizes into finger-like structures. During continuous monolayer transfer the substrate induces a fractional condensation of pure LC-phase, with the fluorescent dye being expelled into the floating monolayer ahead of the three-phase line. Accordingly the monolayer represents an alloy of surfactant plus dye and the substrate-mediated condensation induces the fractional solidification of the pure substance. This perception that the LB-transfer process contains strong analogies to three-dimensional solidification (or melting) opens besides its interest for fundamental research many new material science opportunities like selective doping, zone melting or substrate-induced structuring.

Acknowledgement. We would like to thank those many coworkers who contributed with their detailed work to this overview. Financially, the microfluorescence studies were supported by the Deutsche Forschungsgemeinschaft, the Bundesministerium für Forschung und Technology and by Wacker Chemitronic GmbH, which generously supplied the silicon wafers.

6 References

1. Fukuda K, Sugi M (eds) (1989) Langmuir Blodgett Films 4: Thin Solid Films, vols 178, 179, 180
2. Decher G, Möhwald H, Peterson IR, Riegler H (eds) (1991) Makromol Chem, Macromol Symp 46
3. Special Issue: Organic Thin Films (1991) Adv. Mater 3(1)
4. McConnell H, Tamm LK, Weis M (1984) Proc Natl Acad Sci 81: 3249
5. Lösche M, Möhwald H (1984) Rev Sci Instrum 55: 1968
6. Peters R, Beck K (1983) Proc Natl Acad Sci USA 80: 7183
7. Seul M, Subramaniam S, McConnell H (1985) J Phys Chem 89: 3592
8. Keller DJ, McConnell H, Moy VT (1986) J Phys Chem 90: 2311
9. Subraminam S, McConnell H (1987) J Phys Chem 91: 1715
10. Hirshfeld CL, Seul M (1990) J Physique 51: 1537
11. Seul M, Sammon MJ (1990) Phys Rev Lett 64: 1903
12. Seul M, Sammon MJ, Monar LR (1991) Rev Sci Instr 62: 784
13. Muller P, Gallet F (1991) Phys Rev Lett 67: 1106
14. Muller P, Gallet F (1991) J Phys Chem 95: 3257
15. Qiu X, Ruiz-Garcia J, Stine KJ, Knobler CM (1991) Phys Rev Lett 67: 703
16. Moore B, Knobler CM, Broseta D, Rondelez F (1986) J Chem Soc Faraday Trans 2 82: 1753
17. Lucassen J, Akamatsu S, Rondelez F (1991) J Coll Interface Sci 144: 434
18. To K, Rondelez F (1991) C. R. Acad Sci Paris, t 313, Serie II: 599
19. Möhwald H (1988) Thin Solid Films 159: 1
20. Möhwald H (1988) J Mol Electron 4: 47
21. Möhwald H (1988) In: Carter FL, Siatkowski RE, Wohltjen H (eds) Molecular electronic devices, Elsevier, Amsterdam, p 507
22. Kirstein S, Möhwald H (1991) Makromol Chem Symp 46: 463
23. Ahlers M, Müller W, Reichert A, Ringsdorf H, Venzmer J (1990) Angew Chem Int Ed Engl 29: 1269
24. Henon S, Meunier J (1991) Rev Sci Instrum 62: 936
25. Hönig D, Möbius D (1991) J Phys Chem 95: 4590
26. Riegler JE (1988) Rev Sci Instrum 59: 2220
27. Riegler JE, LeGrange JD (1988) Phys Rev Lett 61: 2492
28. Gaines GL (1966) Insoluble monolayers at liquid-gas interfaces, Wiley, New York
29. Dietrich A, Möhwald H, Rettig W, Brezesinski G (1991) Langmuir 7: 539
30. Lösche M, Duwe H, Möhwald H (1988) J Colloid Interf Sci 126: 432
31. Lösche M, Möhwald H (1989) J Colloid Interf Sci 131: 56
32. Heckl WM, Möhwald H (1986) Ber Bunsenges Phys Chem 90: 1159
33. Heckl WM, Lösche M, Cadenhead DA, Möhwald H (1986) Eur Biophys J 14: 11
34. Heckl WM, Cadenhead DA, Möhwald H (1986) Langmuir 4: 1352
35. Andelman D, Brochard F, Joanny JF (1987) J Chem Phys 86: 3673
36. Vanderlick TK, Möhwald H (1990) J Phys Chem 94: 886
37. Miller A, Möhwald H (1986) Europhys Lett 2: 67
38. Miller A, Knoll W, Möhwald H (1986) Phys Rev Lett 56: 2633
39. Miller A, Möhwald H (1987) J Chem Phys 86: 4258
40. Helm CA, Möhwald H (1988) J Phys Chem 92: 1262
41. Flörsheimer M, Möhwald H (1991) Colloids and Surfaces 55: 173
42. Riegler H, Spratte K (1990) In: Blumen A, Klafter J, Haarer D (eds) Dynamical process in condensed molecular systems. Proceedings of the Emil-Warburg symposium, 23–24 April 1990, Thurnau, World Scientific, Singapore
43. Riegler H, Spratte K (1992) Thin Solid Films 210/211: 9
44. Engel M, Merle HJ, Peterson IR, Riegler H, Steitz R (1991) Ber Bunsenges Phys Chem 95: 1514
45. Spratte K, Riegler H (1991) Makromol Chem, Macromol Symp 46: 113
46. Riegler H, LeGrange JD (1990) Thin Solid Films 185: 335

10. Fluorescence Lifetime Imaging and Application to Ca^{2+} Imaging

Joseph R. Lakowicz[1], Henryk Szmacinski[1], Kazimierz Nowaczyk[3],
Klaus W. Berndt[1] and Michael L. Johnson[2]

[1] University of Maryland, School of Medicine, Center for Fluorescence Spectroscopy, Department of Biological Chemistry, 660 West Redwood Street, Baltimore Maryland 21201, USA
[2] University of Virginia, School of Medicine, Department of Pharmacology, Charlottesville, Virginia 22908, USA
[3] Permanent address: University of Gdańsk, Institute of Experimental Physics, Gdańsk, Poland 80952

Abbreviations

CCD	charge coupled device
DMSS	4-dimethylamino-ω-methylsulfonyl-*trans*-styrene
FD	frequency-domain
FLIM	fluorescence lifetime imaging
9-CA	9-cyanoanthracene
Per	perylene
POPOP	*p*-bis[2-(5-phenyloxazazolyl)]benzene
EGTA	[ethylene bis(oxyethylene-nitrilo)]tetraacetic acid
QUIN-2	2-{[2-bis-(carboxymethyl)-amino-5-methylphenoxy]-methyl}-6-methoxy-8-bis-(carboxymethyl)-aminoquinoline

Introduction

Fluorescence spectroscopy is widely utilized for research in the biosciences [1–8]. These applications have been focused on two divergent disciplines, time-resolved fluorescence and fluorescence microscopy. In time-resolved measurements one takes advantage of the high information content of the time-dependent decays to uncover details about the structure and dynamics of macromolecules [4]. Such measurements are performed almost exclusively using ps laser sources coupled with high speed "single-pixel" photodetectors. While some parallel measurements have been reported, these have been for a linear array detector providing wavelength rather than spatial resolution [9]. In contrast, fluorescence microscopy is most often used to determine the localization (intensity) of the species of interest, usually of proteins or other macromolecules [6, 7]. The acquisition of two-dimensional (2D) fluorescence images is preferentially accomplished with low-speed accumulating detectors [10], which are not capable of quantifying ps–ns fluorescence decays. Consequently, the high information content of time-resolved fluorescence is not usually available for studies of microscopic biological samples. This is particularly disadvantageous when one considers the sensitivity of fluorescence decay times to chemical and environmental factors of interest, such as local pH, cation concentration, oxygen, and polarity, to name a few.

We have now combined measurements of fluorescence lifetimes with 2D imaging, to create images in which the lifetimes are determined at each pixel and the lifetimes are used to create contrast in the images. There have been reports of lifetime measurements at specific locations in microscopic samples [11, 12], and in some cases the lifetime measurements have been raster scanned to create crude lifetime images [13, 14]. Our approach is distinct in that the information on the lifetime at each pixel is obtained simultaneously, which should ultimately result in higher temporal resolution in the sequentially acquired images. While data acquisition is presently slow, we can imagine improved apparatus to acquire the data rapidly, and to display real-time FLIM images. Alternatively, by calculating the difference between two phase-sensitive images at different detector phase angles, one can rapidly compute images in which regions with decay times larger or smaller than the desired value can be suppressed, resulting in visualization of regions of distinct environments.

The concept of FLIM is illustrated in Fig. 10.1. Suppose the sample is composed of two regions, each with an equal intensity of the steady-state fluorescence. Assume further that the lifetime of the probe is several-fold longer in the central region of the object (τ_2), as compared to the outer region (τ_1) The longer lifetime in the central region could be due to the presence of a chemical species, binding of the probe to a macromolecule, or other environmental factors. The intensities of the central and outer regions could be equal due to dye

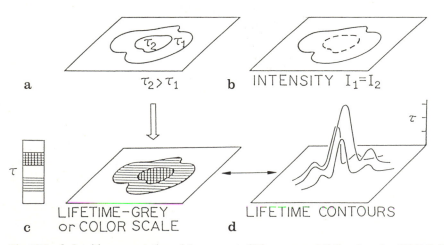

Fig. 10.1a–d. Intuitive presentation of the concept of Fluorescence Lifetime Imaging (FLIM). The object is assumed to have two regions which display the same fluorescence intensity ($I_1 = I_2$) but different decay times, $\tau_2 > \tau_1$. **a** Object; **b** steady state image; **c** grey-scale or color lifetime image; **d** lifetime surface

exclusion or other mechanisms. Observation of the intensity image (upper right) will not reveal the different environments in regions 1 and 2. However, if the lifetimes were measured in these regions then the distinct environments would be detected. The FLIM method allows image contrast to be based on the local decay times, which can be presented on a grey or color scale (Fig 10.1, lower left) or as a 3D surface in which the height represents the local decay times (lower right).

In the present report we describe the FLIM apparatus and the procedure for obtaining the lifetime images. Additionally, we describe two applications of FLIM, these being imaging of fluorophores with known lifetimes, and imaging of calcium concentrations. It is important to recognize that FLIM does not require the probe to display shifts in absorption or emission, such as those which occur in the most widely used Ca^{2+} indicator Fura-2 [15]. Furthermore, fluorescence lifetimes are mostly independent of the probe concentration, at least for the low concentrations and/or thin samples used for microscopy. These two factors suggest that given a suitable fluorophore, the lifetime image can be a chemical image of the chosen species. Hence, there appears to be numerous immediate and potential applications of FLIM to the biological sciences.

2 Creation of Lifetime Images

In order to create lifetime images we perform high-frequency gain-modulation of an image intensifier so as to preserve the decay time information. In our present apparatus we modulate the gain at the same frequency as the light modulation frequency, or at some harmonic of the laser repetition rate. More specifically, we are using the image intensifier as a 2D phase-sensitive detector, in which the signal intensity at each position (r) depends on the phase angle difference between the emission and the gain modulation of the detector. This results in a constant intensity which is proportional to both the concentration of the fluorophore (C) at location r $[C(r)]$, to the cosine of the phase angle difference, to the extent of modulation of the detector (m_D) and to the modulation of the emission at each location $[m(r)]$,

$$I(\theta_D, r) = k\, C(r)[1 + \tfrac{1}{2} m_D\, m(r) \cos\{\theta(r) - \theta_D\}]. \qquad (1)$$

In this expression θ_D is the phase of the gain-modulation, and $\theta(r)$ is the phase angle of the fluorescence. A value of $\theta_D = 0$ results in maximum intensity for a zero lifetime, i.e. scattered light. Our procedure is analogous to the method of phase-sensitive or phase-resolved fluorescence [16, 17]. However, these earlier measurements of phase-sensitive fluorescence were performed electronically on the heterodyne low-frequency cross-correlation signal, whereas our present homodyne measurements are performed electro-optically on the high frequency modulated emission. The phase angle of the fluorophore is related to the apparent phase lifetime τ_θ and the modulation frequency (ω in radians/s) by

$$\tan \theta(r) = \omega\, \tau_\theta(r). \qquad (2)$$

It is not possible to calculate the lifetime from a single phase-sensitive intensity. However, the phase of the emission can be determined by examination of the detector phase angle dependence of the emission, which is easily accomplished by a series of electronic delays in the gain modulation signal or by optical delays in the modulated excitation. The phase-sensitive images can also be used to compute the modulation at each pixel [$m(r)$]. The modulation is related to the modulation life time $\tau_m(r)$ by

$$m(r) = \frac{1}{\sqrt{1 + \omega^2 \tau_m^2(r)}}. \tag{3}$$

The desired information is thus obtained by varying θ_D (Eq. 1), which in turn allows determination of $\theta(r)$ or $m(r)$. In our apparatus we collect a series of phase-sensitive images, in which θ_D is varied over 360 degrees or more. The phase-sensitive intensities at each pixel are used to determine the phase and modulation at each pixel, resulting in the phase angle, modulation or lifetime images. This procedure is illustrated schematically in Fig. 10.2. The object is illuminated with intensity modulated light. A gain-modulated image intensifier, and a CCD camera are used to obtain a series of phase-sensitive images. In these images, the phase-sensitive intensity for each region is expected to vary as the cosine of the phase angle difference (Eq. 1). Also, the modulation of the image is expected to depend on the lifetime in each region according to Eq. (3).

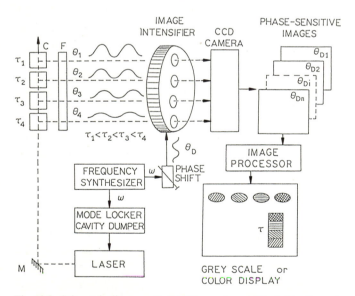

Fig. 10.2. Schematic diagram of a FLIM experiment. The "object" consists of a row of four cuvettes, and has regions with different decay times, τ_1 to τ_4. This object is illuminated with intensity modulated light. The spatially-varying emission is detected with a phase-sensitive image intensifier, which is imaged onto a CCD camera. The laser system is a cavity-dumped dye laser, which is synchronously pumped by a mode-locked and frequency-doubled NdYAG laser

The data sets for FLIM are rather large (512×512 pixels, and 520 kbyte storage for each image), which can result in time-consuming data storage, retrieval and processing. In order to allow rapid calculation, we developed an algorithm to calculate the phase and modulation images which uses each image one time, and only one image is in memory at any given time. This algorithm (called CCDFT) is described elsewhere [18]. In order to test this algorithm we examined the dependence of the phase-sensitive intensities on the detector phase angle. For this analysis we used an average intensity from the central illuminated spot consisting of approximately 5 by 10 pixels. These intensities were fit to Equation 1 (using a program called Cosine) and used to recover the phase and modulation of the emission. We note that all the phase angles and modulations [$\theta(r)$ and $m(r)$] initially contain shifts and amplitude changes due to instrumental factors. In the present report all detector phase angles are given as instrumental values $\theta'_D = \theta_D + \theta_I$, where θ_I is a phase shift intrinsic to the apparatus and the value required to give the correct phase angle for the reference sample. Similarly, the apparent modulation of the image is "normalized" so the modulation of the standard matches the expected value.

3 Instrumentation for FLIM

A detailed description of the FLIM apparatus can be found elsewhere [19]. The light source is a ps dye laser system, consisting of a mode-locked Antares NdYAG, which synchronously pumps the cavity-dumped dye lasers containing either rhodamine 6G or pyridine 1. The pulse repetition rate was varied from 3.81 to 76.2 MHz yielding similar results, and we either used the fundamental or higher harmonics of the pulse train.

The detector was a CCD camera from Photometrics (series 200) with a thermoelectrically cooled PM-512 CCD. The gated image intensifier (Varo 510-5772-310) was positioned between the target and the CCD camera. The intensifier gain was modulated by a RF signal applied between the photocathode and microchannel plate (MCP) input surface. This signal was capacitively isolated from the high voltage across the intensifier. Phase delays were introduced into this gating signal using calibrated coaxial cables.

The target consisted of rows of cuvettes, each containing scatterers, standard fluorophores, or fluorophores with environment-sensitive lifetimes (Fig. 10.2). The laser passed through the center of the cuvettes. For calculations of the FLIM images we used one of the cuvettes containing a lifetime standard of known phase and modulation. For FLIM of the standards, DMSS was taken as the standard with $\tau = 40$ ps [20]. For FLIM of Quin-2 the sample containing the most calcium was typically taken as the reference. The phase and modulation of the other cuvettes were calculated using this reference, as described previously [21].

4 Experimental Methods

Lifetimes recovered from the FLIM measurements were compared with those obtained using standard frequency-domain(FD)measurements and instrumentation [22–24]. For the FLIM measurements polarizers were not used to eliminate the effects of Brownian rotation, which are most probably insignificant for these decay times and viscosities. The standard fluorophores (POPOP, perylene and 9-CA) were dissolved in air-equilibrated cyclohexane. The reference fluorophore DMSS was dissolved in methanol. Quin-2 was obtained from Sigma (Q-4750, lot 110H0366, FW 541.5) and used without further purification. Ca^{2+} concentrations were obtained using the Calibrated Calcium Buffer Kit II, obtained from Molecular Probes, Eugene, Oregon. All $[Ca^{2+}]$ refer to the concentration of free calcium. For all measurements the temperature was 20°C, except for the FLIM measurements, which were done at room temperature near 25°C. The excitation wavelength was 342 nm for Quin-2 and 355 nm for the standard fluorophores. The emission was observed through a Corning 3-72 filter for Quin-2 and 3-75 for the standards. The recommended excitation wavelength for Quin-2 is 339 nm [25]. We did not use the precise value because of inadequate output of our frequency-doubled pyridine-1 dye laser at this wavelength.

5 Results

5.1 FLIM of Lifetime Standards

Since the technique of FLIM is new, it is necessary to test objects with known lifetimes. Hence, we imaged rows of cuvettes, each containing a different fluorophore with a different lifetime. This approach has the advantage of allowing comparison of the FLIM results with the standard frequency-domain (FD) lifetime measurements of the same samples. We used four fluorophores with known lifetimes, these being a styrene derivative (DMSS) with a 40 ps lifetime in methanol [20], and POPOP (1.10 ns), perylene (3.75 ns) and 9-cyanoanthracene (9-CA, 9.90 ns), all dissolved in air-equilibrated cyclohexane. The emission spectra are shown in Fig. 10.3. These spectra overlap nearly completely on the wavelength scale, which was deliberately chosen in order to emphasize that the samples were being distinguished on the basis of the lifetimes. The lifetimes of these fluorophores result in phase shifts of 0.7, 18.9, 49.1 and 71.7 degrees at 49.335 MHz as measured on the FD instrument (Table 1). The pulse repetition rate was 3.81 MHz, but the measurement frequency of 49.53 MHz is determined by the frequency applied to the image intensifier. In a number of experiments we used pulse repetition rates ranging from 3.81 to 76.2 MHz, and we found no changes in the phase angle or modulation, as expected for a harmonic content FD measurement.

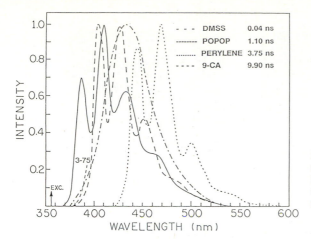

Fig. 10.3. Emission spectra of standard fluorophores used for FLIM. The emission was observed through a corning 3-75 filter. Excitation was at 355 nm.

Table 10.1. Phase and modulation data for the standard fluorophores[a] at 49.53 MHz

Compound	method[b]	Phase		Modulation	
		θ (deg)	τ_θ (ns)[d]	m	τ_m (ns)[d]
DMSS	FD	0.7	0.04	1.0	0.04
	Cosine[c]	$\langle 0.7 \rangle \pm 4.7$	0 to 0.30	$\langle 1.0 \rangle \pm 0.09$	0 to 1.27
	CCDFT[c]	$\langle 0.7 \rangle \pm 1.0$	0.04 ± 0.04	$\langle 1.0 \rangle \pm 0.02$	0 to 0.65
POPOP	FD	18.9	1.10	0.946	1.10
	Cosine	15.4 ± 4.2	0.89 ± 0.25	0.994 ± 0.07	0.35 (0 to 1.33)
	CCDFT	16.6 ± 0.5	0.95 ± 0.05	0.928 ± 0.02	1.29 ± 0.10
Perylene	FD	49.1	3.70	0.645	3.80
	Cosine	48.3 ± 3.9	3.61 ± 0.45	0.797 ± 0.06	2.44 ± 0.45
	CCDFT	49.0 ± 0.5	3.68 ± 0.05	0.691 ± 0.06	3.36 ± 0.10
9-CA	FD	71.7	9.70	0.303	10.1
	Cosine	77.7 ± 3.2	14.7 ± 4.2	0.383 ± 0.02	7.75 ± 0.5
	CCDFT	78.5 ± 0.5	14.7 ± 0.50	0.309 ± 0.01	9.90 ± 0.4

[a] Excitation wavelength 355 nm, corning 3-75 emission filter, 25°C.
[b] FD, standard frequency-domain measurements at 49.335 MHz; Cosine, fit of the averaged phase-sensitive intensities, measured at 49.53 MHz; to Eq. 1; CCDFT, pixel-by-pixel values [18].
[c] The angular brackets indicate these values were used as reference values.
[d] The uncertainty ranges of τ_θ and τ_m are not symmetrical, and an average values are shown. For the Cosine method the uncertainties were from the usual assumptions in a least-squares fit. For the CCDFT calculations the uncertainties in the standard value represent the standard deviations of about 120 pixels (6 × 20 pixels) from the central region of the illuminated spot.

The primary information for FLIM is presented in Fig. 10.4, which shows the phase-sensitive images of the illuminated cuvettes. One notices that both the absolute and relative intensity of the illuminated spots vary with the detector phase angle. These phase-sensitive intensities were used to compute the phase angles, modulation and apparent lifetimes (Table 10.1). These values are in good agreement with those obtained from the standard FD measurements.

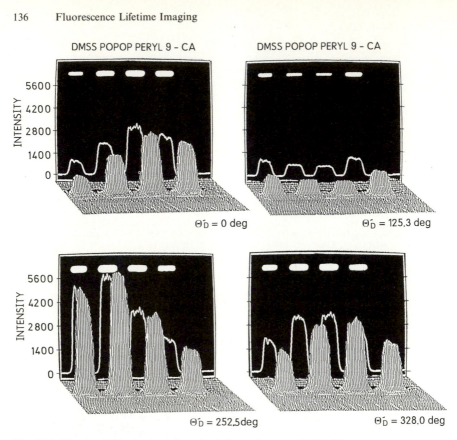

Fig. 10.4. Phase-sensitive images of standard fluorophores at 49.53 MHz

It is somewhat difficult to see the phase differences between the cuvettes in the phase-sensitive images (Fig. 10.4). However, the relative phase angles are easily visualized by a plot of the averaged phase-sensitive intensity versus detector phase angle (Fig. 10.5). For this figure we used an average phase-sensitive intensity obtained for approximately 5 by 10 pixels, using the same spatial window for all data files. The phase angle of the fluorescence sample can be determined from the phase shift relative to the reference. Inspection of this figure reveals that the phase sensitive intensities of the fluorophores are delayed in phase relative to the 40 ps reference, and that the phase shift increases and the modulation decreases in the order DMSS, POPOP, perylene, 9-CA. Also, the phase angles are larger, and the modulations smaller, at higher frequencies (not shown). These phase angles and modulations are in good agreement with those observed from the direct FD measurements (Table 10.1). These results suggest that the FLIM apparatus is providing a reliable measure of the fluorescence phase angle and modulation.

The phase-sensitive images (Fig. 10.4) were used to calculate the phase (Fig. 10.6, left) and modulation images (Fig. 10.6, right), using the algorithm described

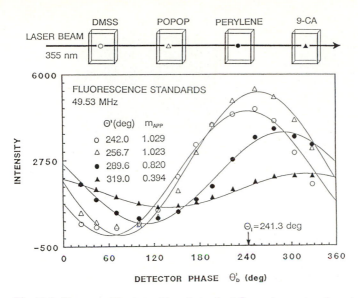

Fig. 10.5. Phase-sensitive intensities of standard fluorophores at various detector phase angles

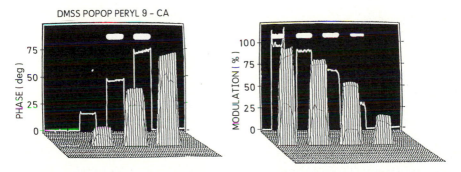

Fig. 10.6. Phase angle (*left*) and modulation (*right*) of the standards observed at 49.53 MHz

elsewhere [18]. Only those pixels with phase-sensitive intensities higher than 250 (about 5% of the maximum intensity) were used to compute the phase and modulation images. For reasons we do not presently understand, elimination of these low intensity values from the computation appears to improve the agreement between the phase and modulation values obtained from the FLIM apparatus, and those found from the FD measurements or the Cosine fits. Also, for these images we only displayed the phase and modulation for regions of the image where the stationary intensity was 5% or greater of the peak intensity in the steady-state intensity image. One notices that the phase angles increase (left to right) and the modulation decreases, as expected for the increasing lifetimes of the standards.

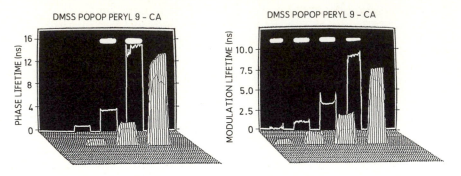

Fig. 10.7. Phase and modulation lifetime image of the standards observed at 49.53 MHz

Once the phase angle and/or modulation images are available they can be subjected to any manner of image processing. For instance, the phase and modulation lifetime image is simply a numerical conversion of the phase and modulation image using Eqs. (2) and (3), respectively. The phase lifetime image of the standards at 49.53 MHz is shown in Fig. 10.7 (left). The lifetimes recovered from this image are in good agreement with those obtained from the FD measurements, and from fitting the phase-sensitive intensities (Fig. 10.5 and Table 10.1). Any desired range of phase angles or lifetimes can be chosen as the basis for contrast, and pseudo 3D projections are possible. Figure 10.7 (right) shows the modulation lifetime image at 49.53 MHz. Agreement between these two methods suggests both the phase and modulation images can be used in parallel. Average or composite images of phase and modulation lifetimes may allow one to increase the robustness of the method. For instance, the modulation data may be more reliable for lifetime measurements in regions where the lifetime-modulation frequency product, $\omega\tau$, is large. Conversely, the phase angle images may be more reliable for measurement in short decay time regions of the image, where the modulation lifetimes are less reliable. In Figure 10.7 we chose to compute lifetimes only where the intensity was 5% or higher than the peak intensity of the steady state intensity image (with no gain modulation). The lack of random noise down to the 5% limit suggests that lifetime imaging can be accomplished over a 20-fold range of intensities. This will be valuable in fluorescence microscopy in that one cannot force the probe concentration and/or intensity to be equal in all regions of a cell or microscopic specimen. Further refinements in the FLIM methodology may further expand the dynamic range for the local intensities.

5.2 FLIM of Quin-2

Measurement of the intracellular concentrations of Ca^{2+} is of interest for understanding its role as a second messenger, and the response of cells to various stimuli. Equally important is the imaging of the local calcium concentration to study cell function. At present, most measurements of intracellular calcium or Ca^{2+} imaging are performed using fluorescence indicators [25, 26]. These dyes (Quin-2 and Fura-2) change intensity in response to Ca^{2+}. The second generation dye Fura-2 is currently preferred for a variety of chemical and biochemical reasons, as discussed elsewhere [15, 28]. The primary advantage of Fura-2 over Quin-2 appears to be the shift in its excitation wavelength in response to Ca^{2+} [29–31], which allows calculation of the calcium concentration from the ratio of fluorescence intensities at two excitation wavelengths, thus providing a measure of the $[Ca^{2+}]$ which is independent of the probe concentration. While Quin-2 can also be used as a ratiometric probe, this use is not favored due to the need for 340 nm excitation, and its weak absorption at longer wavelengths. Nonetheless, Quin-2 has a number of advantages for measurements of Ca^{2+}, such as minimal interference for Mg^{2+}, a known stoichiometry for Ca^{2+}, and a favorable dissociation constant [25].

In order to obtain lifetime images of Ca^{2+} it is necessary for the probe to display a change in lifetime in response to Ca^{2+}. Absorption and emission spectra of Quin-2 are shown in Figs. 10.8 and 10.9, respectively. These spectra agree with previously published spectral data for Quin-2 [25, 26], and show that our sample of Quin-2 displayed the expected dependence on Ca^{2+}. Also, the intensity increases 4.6-fold upon complexation with Ca^{2+}, which is within the expected range [25, 26]. Based on the increased yield, one might expect the lifetime of Quin-2 to increase on binding Ca^{2+}. However, an increased lifetime cannot be predicted with certainty because of the unknown static versus dynamic quenching processes operative in Quin-2 and its Ca^{2+} complexes.

Frequency-domain lifetime data for Quin-2 with various amounts of Ca^{2+} are shown in Fig. 10.10. One notices that the frequency-response shifts to lower frequencies with increasing amounts of Ca^{2+}, indicating that the mean lifetime

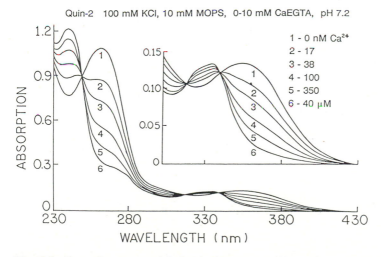

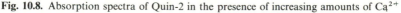

Fig. 10.8. Absorption spectra of Quin-2 in the presence of increasing amounts of Ca^{2+}

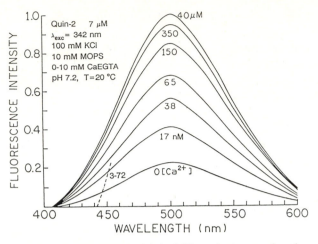

Fig. 10.9. Emission spectra of Quin-2. The excitation wavelength was 342 nm. The *dashed line* shows the transmission of the Corning 3-72 emission filter used to isolate the emission during the FLIM or FD measurements

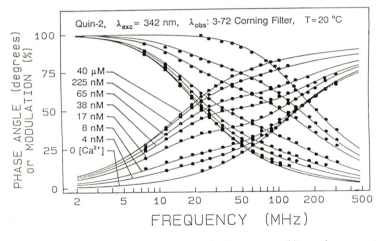

Fig. 10.10. Frequency-response of Quin-2 in the presence of increasing amount of Ca^{2+}

is increasing. At intermediate concentrations of Ca^{2+} the shape of the frequency-response is more complex than at the extreme of low and high Ca^{2+} concentrations. This is because the Quin-2 decay in the free and Ca^{2+}-bound states is dominantly a single exponential, and the decay becomes doubly exponential under conditions of partial Ca^{2+} saturation where the emission from both species contribute to the measured phase and modulation data. This increase in the lifetime of Quin-2 with Ca^{2+} binding is in agreement with a recent report [27].

It is of interest to examine the Ca^{2+}-dependent phase and modulation data at selected frequencies, in order to select an optimal frequency consistent with the lifetimes displayed by the samples and the useful frequency range of the instrumentation. Phase and modulation data for Quin-2 at selected frequencies are shown in Fig. 10.11. Substantial Ca^{2+}-dependent phase angle changes are seen from 34 to 72 MHz. Somewhat smaller changes in phase and modulation are seen at lower (16 MHz) and higher (104 MHz) frequencies (Fig. 10.10), but these changes are more than adequate with the current precision of our FLIM instrumentation. Conveniently, this range is consistent with that of our FLIM instrument, which operates to about 150 MHz. For the subsequent FLIM experiments we selected 49.335 MHz, which is in the center of the Ca^{2+} response curve and provides maximal changes in phase and modulation (Fig. 10.11).

The FLIM images are calculated from a series of phase-sensitive images (not shown), using the algorithm described in [18]. This image can be presented as a phase angle image, or transformed into phase lifetimes using Eq. (2). In Fig. 10.12, the height of the surface is the phase angle (left) or phase lifetime

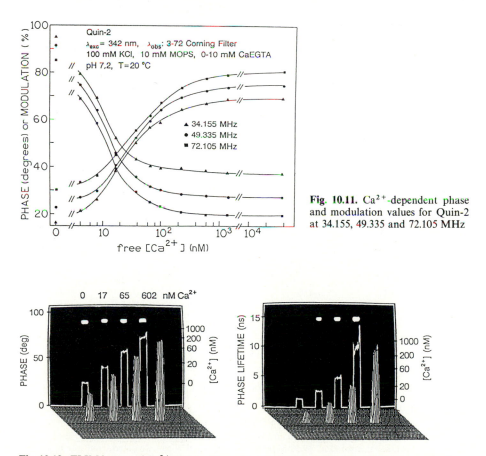

Fig. 10.11. Ca^{2+}-dependent phase and modulation values for Quin-2 at 34.155, 49.335 and 72.105 MHz

Fig. 10.12. FLIM images of Ca^{2+} obtained from the phase angle image. The calcium concentrations were obtained from the calibration curve in Fig. 10.11

(right). Also shown is a line (on the background panels) representing the phase angles drawn through the center of the four cuvettes. The phase angles increase with increasing concentrations of Ca^{2+}. The phase lifetimes obtained from the FLIM measurements are in excellent agreement with those measured using the standard FD instrumentation except for the sample without Ca^{2+} (Table 10.2). Calculation of a Ca^{2+} concentration image is a simple transform of these data according to the calibration curve in Fig. 10.11. Lifetime and/or Ca^{2+}-images can also be calculated from the modulation at each pixel (Fig. 10.13). Notice that the modulation decreases with increasing $[Ca^{2+}]$, whereas in Fig. 10.9 the phase angles increase with increasing $[Ca^{2+}]$. However, in both cases the

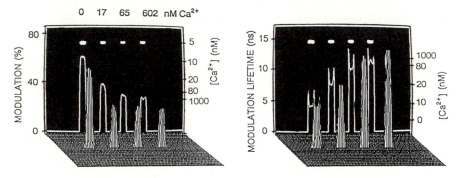

Fig. 10.13. FLIM images of Ca^{2+} obtained from the modulation image. The modulation lifetimes were obtained from the calibration curve in Fig. 10.11

Table 10.2. Phase and Modulation of Quin-2[a]

$[Ca^{2+}]$ (nM)	Method[b]	Phase		Modulation	
		θ (deg)	τ_θ (ns)	m	τ_m (ns)
0	FD	21.6	1.27	0.921	1.36
	Cosine	24.3 ± 6.0	1.45 ± 0.40	0.639 ± 0.08	3.87 ± 0.80
	CCDFT	24.5 ± 4.0	1.47 ± 0.27	0.611 ± 0.09	4.16 ± 1.00
17	FD	40.6	2.77	0.447	6.46
	Cosine	40.4 ± 8.1	2.74 ± 0.80	0.408 ± 0.07	7.19 ± 1.9
	CCDFT	40.6 ± 6.0	2.76 ± 0.60	0.391 ± 0.08	7.67 ± 1.9
65	FD	59.1	5.39	0.314	9.76
	Cosine	57.2 ± 6.1	4.99 ± 1.2	0.320 ± 0.04	9.52 ± 1.3
	CCDFT	57.3 ± 7.0	5.01 ± 1.4	0.310 ± 0.04	9.86 ± 1.4
602	FD	72.3	10.11	0.271	11.46
	Cosine	⟨72.3⟩ ± 5.7	10.11 ± 3.8	⟨0.27⟩ ± 0.03	11.46 ± 1.5
	CCDFT	⟨72.3⟩ ± 7.0	10.11 ± 5.0	⟨0.27⟩ ± 0.04	11.46 ± 1.8

[a] Excitation wavelength 342 nm, Corning 3-72, emission filter, a 25°C at 49.53 MHz.
[b] FD, standard frequency-domain measurements at 49.335 MHz; Cosine fit to averaged phase-sensitive intensities, measured at 49.53 MHz to Eq. 1; CCDFT as calculated from our algorithm [18].
⟨ ⟩ Used as reference values.

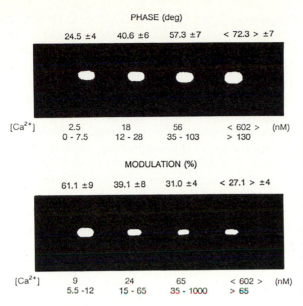

PHASE (deg)

| 24.5 ±4 | 40.6 ±6 | 57.3 ±7 | < 72.3 > ±7 |

| [Ca²⁺] | 2.5 | 18 | 56 | < 602 > | (nM) |
| | 0 - 7.5 | 12 - 28 | 35 - 103 | > 130 | |

MODULATION (%)

| 61.1 ±9 | 39.1 ±8 | 31.0 ±4 | < 27.1 > ±4 |

| [Ca²⁺] | 9 | 24 | 65 | < 602 > | (nM) |
| | 5.5 -12 | 15 - 65 | 35 - 1000 | > 65 | |

Fig. 10.14. Ca^{2+} imaging using a grey-scale. The Ca^{2+} is obtained from the phase angle and modulation calibration curves (Fig. 10.11). The values above the figures are the phase and modulation values from the pixel-by-pixel CCDFT analysis, with the uncertainties representing the average variation within the illuminated spot. Because the phase and modulation are not linearly dependent on the $[Ca^{2+}]$, the uncertainties in $[Ca^{2+}]$ are not symmetrical about the mean value

apparent lifetimes increase with calcium concentration, and these lifetimes are in agreement with the expected values (Table 10.2).

Examination of Figs. 10.12 and 10.13 reveals peaks on the sides of the lifetime surfaces (right), or equivalently rounded edges on the modulation surface (left). These structures are surprising because each cuvette is a homogeneous solution and is expected to display a single phase angle or modulation value. We are currently investigating this phenomena, which presently appears to be the result of a computational effect rather than an electro-optic phenomena in the FLIM apparatus. More specifically, these peaks appear to be due to the use of data files with near zero intensity.

An alternative presentation of the Ca^{2+} images is shown in Fig. 10.14. The local calcium concentrations can be determined from the grey-scale intensity. Such images may be most appropriate for Ca^{2+} imaging of cells, particularly if colors are assigned for each range of calcium concentration, resulting in easier visualization of the Ca^{2+} concentration image.

5.3 Phase Suppression Imaging of Ca^{2+}

A unique property of FLIM is the ability to suppress the emission for any desired lifetime and/or calcium concentration. Suppression of the emission with any given decay time can be accomplished by taking the difference on two phase sensitive images obtained for detector phase angles of θ_D and $\theta_D + \Delta$. In the difference image $\Delta I = I(\theta_D + \Delta) - I(\theta_D)$ components are suppressed ($\Delta I = 0$)

with a phase angle θ_S which is given by

$$\theta_S = \theta_D + \Delta/2 \pm n \cdot 180. \qquad (4)$$

Regions of the image with a decay time of $\tau_S = \omega^{-1} \tan \theta_S$ have an intensity of zero in the difference image. This concept is shown schematically in Fig. 10.12 for $\Delta = 180°$. Components with a lifetime τ_2 larger than the suppressed lifetime appear negative in the difference image ($\Delta I_2 < 0$) and components with a shorter lifetime are positive ($\Delta I_1 > 0$). This relationship is reversed if one calculates $I(\theta_D) - I(\theta_D + \Delta)$. A more complete description of this suppression method will be presented elsewhere [18].

The use of difference images to suppress the emission for various concentrations of Ca^{2+} is shown in Fig. 10.16. In the left panel we chose to suppress the emission from areas with $[Ca^{2+}] \geqslant 80$ nM. If one examines only the grey-scale representation, and sets negative intensities to the background color, then only

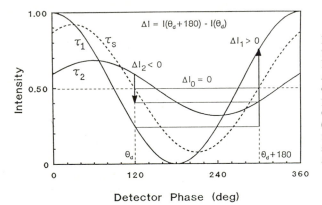

Fig. 10.15. Intuitive descriptions of phase suppression. In a difference image with $\Delta O = I(\theta_D + 180) - I(\theta_D)$ a component with $\theta = \theta_D$ is completely suppressed. Components with longer lifetimes (phase angles) appear to be negative, and those with shorter lifetimes (phase angles) appear to be positive

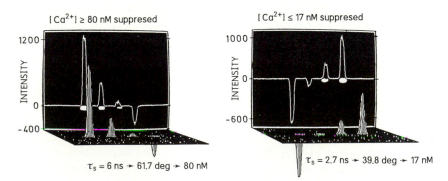

Fig. 10.16. Ca^{2+} images with suppression of regions with $[Ca^{2+}] \geqslant 80$ nm (*left*) and $Ca^{2+} \leqslant 17$ nM (*right*). The left and right suppression images were calculated from the phase sensitive images $I(\theta_D)$ using $I(348.2°) - I(152.4°)$ and $I(152.4°) - I(304.6°)$, respectively

regions with $[Ca^{2+}] \leqslant 80$ nM are observed. Remarkably, the sample with 65 nM Ca^{2+} still shows positive intensity in the difference image. Alternatively, one can suppress the emission from regions with $[Ca^{2+}] \leqslant 17$ nM (right). In this case, the grey-scale image only shows regions with $[Ca^{2+}] \geqslant 17$ nM. The samples with 0 and 17 nM Ca^{2+} both show negative intensities in the difference image, or no intensity in the grey-scale representation. This ability to selectively visualize regions with high or low Ca^{2+} may be useful in evaluation of the role of Ca^{2+} in the control of cellular processes. Acquisition and computation of complete FLIM images is presently time consuming. In contrast, phase suppression images require only the difference of two images without further numerical analysis, making it easier to acquire and display real time images.

6 Discussion

Fluorescence lifetime imaging provides a new opportunity for the use of fluorescence in cell biology. This is because the lifetimes of probes can be sensitive to a variety of chemical or physical properties, many of which are of interest for studies of intracellular chemistry and physiology. An advantage of FLIM is the insensitivity of lifetime measurements to the local probe concentration and photobleaching. Consequently, one does not require dual-wavelength ratiometric probes. Instead, one needs a change in lifetime, which may occur in any Ca^{2+} probe which changes intensity in response to Ca^{2+}.

At present, the selection of fluorophores for FLIM is not straightforward. This is because most sensing work does not rely on lifetimes and the probe lifetimes are often unknown. For instance, the Ca^{2+} probes Fura-2 [32] and Indo-1 (Lakowicz et al., unpublished observation) showed only small changes in phase angles in response to Ca^{2+}. Nonetheless, one can expect lifetime probes to become available as the available sensors are tested. It should be noted that it may be easier to obtain lifetime probes for pH, Cl^-, Na^+ and K^+ than wavelengths shifting probes. For instance, it is known that Cl^- is a collisional quencher and alters the lifetime of quinine [33]. Hence, the chloride intensity indicator SPQ (6-methoxy-N-(3-sulfopropyl) quinolinium) is probably also a lifetime probe for Cl^-, as suggested by Tsien [34], and confirmed by the experiments of Illsley and Verkman [35]. Hence, elimination of the requirement for dual-wavelength excitation and/or emission, may result in the rapid introduction of many FLIM probes.

In spite of the current lack of a library of FLIM probes, one can expect many probes to be identified. More specifically, the fluorescence lifetimes of probes can be altered by oxygen [36–38], pH [39–41], energy transfer [42–44], as well as a variety of other factors and/or quenchers [45–49]. It will also be of interest to examine lifetime images of stained chromosomes [50], where the local G–C/A–T ratios may affect the decay times of the stains [51]. Such imaging could rely on the wealth of knowledge of the fluorescent properties of dye–DNA complexes [52]. One can also envisage imaging of free versus protein-bound

NADH in cells, recovered from the 0.5 and 1.5 ns lifetimes of free and bound NADH, respectively [53].

It should be noted that the apparatus required for FLIM is a modestly straightforward extension of that already in use in fluorescence microscopy. Slow-scan CCD cameras are in use, as are laser sources. The image intensifier is commercially available, and is easily gain-modulated with low voltages. Phase angle or lifetime image files are easily rewritten in the format of the image processing software packages, so that these powerful image manipulation programs remain available after collection and processing of the lifetime images.

It should also be noted that the FLIM measurements avoid a difficult problem in quantitative microscopy. Ratiometric imaging requires the overlap of the images observed at two emission wavelengths. These images may be different due to incomplete color correction of the optical elements. Since the FLIM measurement provides the equivalent information at a single wavelength, image registration is no longer a problem. Additionally, dynamic information may be obtained directly from the time-dependent intensity decays. For instance, the use of time-dependent donor decays would allow imaging based on energy transfer, i.e. proximity imaging, without the need to measure both the donor and the acceptor. That is, the donor decay time will be characteristic of the presence of a nearby acceptor, independent of the local concentration of the donor.

In conclusion, FLIM offers new opportunities for chemical imaging of cellular systems. Of course, much additional work is needed to realize this potential.

Acknowledgements. The authors acknowledge support from grants from the National Science Foundation (DIR-8710401 and DMB-8804931, Center for Fluorescence Spectroscopy and Institutional grants), and for support from the Medical Biotechnology Center and Graduate School at the University of Maryland, without whose support these experiments could not have been accomplished.

7 References

1. Dewey TG (ed) (1991) Biophysical and biochemical aspects of fluorescence spectroscopy. Plenum, New York
2. Jameson DM, Reinhart G (1989) Fluorescent biomolecules. Plenum, New York
3. Lakowicz JR (1983) Principles of fluorescence spectroscopy. Plenum Press, New York, 496 pp
4. Lakowicz JR (ed) (1990) Time-Resolved laser spectroscopy in biochemistry II. SPIE Press, Washington, DC, 850 pp
5. Demchenko AP (1986) Ultraviolet spectroscopy of proteins. Springer, Berlin Heidelberg, New York
6. Wang Y, Taylor DL (eds) (1989) Fluorescence microscopy of living cells in culture, Part A: Fluorescent Analogs, Labeling Cells, and Basic Microscopy. Academic, New York, 503 pp
7. Taylor DL, Wang Y (eds) (1989) Fluorescence microscopy of living cells in culture, Part B: Quantitative Analogs, Microscopy – Imaging and Spectroscopy. Academic, New York
8. Inoué S (1986) Video microscopy. Plenum, New York, 584 pp
9. Gratton E, Feddersen B, vandeVen M (1990). Parallel acquisition of fluorescence decay using array detectors. In: Lakowicz JR (ed) Time-resolved laser spectroscopy in Biochemistry II. Proc of SPIE, 1204: 21

10. Hiraoka Y, Sedat JW, Agard DA (1987) Science. 23: 36
11. Keating SM, Wensel TG (1990). Nanosecond fluorescence microscopy of single cells. In: Lakowicz JR (ed) Time-resolved laser spectroscopy in biochemistry II; Proc. of SPIE, 1204: 42–48.
12. Arndt-Jovin DJ, Latt SA, Striker G, Jovin TM (1979) J of Histochem and Cytochem 27: 87
13. Wang SF, Kitajima S, Uchida T, Coleman DM, Minami S (1990). Applied Spectroscopy. 44: 25
14. Wang XF, Uchida T, Minami S (1989) Applied Spectroscopy. 43: 840
15. Grynkiewicz G, Poenie M, Tsien RY (1985) J of Biol Chem 260: 3440
16. Lakowicz JR, Cherek HC (1981) Journal of Biochemical and Biophysical Methods 5: 19
17. Lakowicz JR, Cherek HC (1981) J Biol Chemistry 256: 6348
18. Lakowicz JR, Szmacinski H, Nowaczyk K, Berndt K, Johnson ML (1992) Fluorescence lifetime imaging. Biophysical Journal 202: 316.
19. Lakowicz JR, Berndt KW (1991) Rev of Sci Instr 67: 1727
20. Gryczynski I, Lakowicz JR unpublished observation.
21. Lakowicz JR, Cherek H, Balter A (1981) Journal of Biochemical and Biophysical Methods 5: 131
22. Lakowicz JR, Maliwal BP (1985) Biophysical Chemistry 21: 61
23. Lakowicz JR, Laczko G, Gryczynski I (1986) Reviews of Scientific Instrumentation 57: 2499
24. Laczko G, Lakowicz JR, Gryczynski I, Gryczynski Z, Malak H (1990) Reviews of Scientific Instruments 61: 2331
25. Tsien R, Pozzan T (1989) Methods of Enzymology 172: 230
26. Tsien R (1980) Biochem 19: 2396
27. Miyoshi N, Hara K, Kimura S, Nakanishi K, Fukuda M (1991) Photochemistry and Photobiology 53: 415
28. Komada H, Nakabayashi H, Nakano H, Hara M, Yoshida T, Takanari H, Izutsu K (1989) Cell Structure and Function 14: 141
29. Moore ED, Becker PL, Fogarty KE, Williams DA, Fay FS (1990) Cell Calcium 11: 157
30. Roe MW, Lemasters JJ, Herman B (1990) Cell Calcium 11: 63
31. Goldman WF, Bova S, Blaustein MP (1990) Cell Calcium 11: 221
32. Keating SM, Wensel TG (1991) Biophys J 59: 186
33. Chen RF (1974) Anal Biochemistry 57: 593
34. Tsien RY (1989) Methods in Cell Biology 30: 127
35. Illsley NP, Verkman AS (1987) Biochemistry 26: 1215
36. Kautsky H (1930) Trans Faraday Soc 35: 262
37. Lakowicz JR, Joshi NB, Johnson ML, Szmacinski H, Gryczynski I (1987) Journal of Biological Chemistry 262: 10907
38. Lakowicz JR, Weber G (1973) Biochemistry 12: 4161
39. Laws JR and Brand L (1979) J Phys Chem 83: 795
40. Gafni A and Brand L (1978) Chem Phys Lett 58: 346
41. Jameson DM, Weber G (1981) The Journal of Physical Chemistry 85: 95
42. Förster T (1948) Ann Phys (Leipzig). 2: 55 (Translated by Knox RS)
43. Stryer L (1978) Ann Rev Biochem 47: 819
44. Steinberg IZ (1971) Ann Rev Biochem 40: 83
45. Eftink MR, Ghiron C (1981) Anal Biochem 114: 199
46. Lehrer SS (1971) Biochemistry 10: 3254
47. Ware WR, Watt D, Holmes JD (1974) J Amer Chem Soc 96: 7853
48. Leto TL, Roseman MA, Holloway PW (1980) Biochemistry 19: 1911
49. Thulborn KR, Sawyer WH (1978) Biochim Biophys Acta 511: 125
50. Weisblum B, de Haseth PL (1971) Proc Nat Acad Sci USA 69: 629
51. Arndt-Jovin D, Latt SA, Striker G, Jovin TM (1979) The Journal of Histochemistry and Cytochemistry 27: 87
52. Steiner RF, Kubota Y (1983) In: Fluorescent Dye-Nucleic Acid Complexes. Excited States of Biopolymers. Plenum Press, New York, p 203
53. Scott TG, Spencer RD, Leonard NJ, Weber G (1970) J Amer Chem Soc 92: 687

11. Ether Phospholipids in Membranes: Applications of Phase and Steady-State Fluorometry

A. Hermetter*, E. Prenner, J. Loidl, E. Kalb, A. Sommer and F. Paltauf

Institut für Biochemie und Lebensmittelchemie, Technische Universität Graz, Petersgasse 12/II, A-8010 Graz, Austria

Almost all procaryotic and eucaryotic cellular membranes contain diacyl glycerophospholipids (Fig. 11.1) as their predominant bilayer components. Most animal cells exhibit in addition, large proportions of ether lipids in their membranes [1]. They belong predominantely to the plasmalogen type. They are alkenylacylglycerophospholipids in which the hydrophobic alkyl chain is linked to position 1 of glycerol via an enolether bond (Fig. 11.1). Increasing evidence has already been accumulated that alkenylacyl and diacyl glycerophospholipids exhibit different properties in artificial as well as in natural bilayer systems [2, 3].

New aspects of ether phospholipids in artificial and biological membranes are now being investigated by means of different fluorescence methods. We synthesized alkenylacyl, alkylacyl, and diacyl glycerophospholipids of the choline and ethanolamine type, containing a fluorescent acyl chain covalently bound to position 2 of glycerol (Fig. 11.2). The fluorescent substituents were pyrene decanoyl, parinaroyl, diphenylhexatrienyl (DPH) propionyl, and nitro-benzoxadiazolyl (NBD) dodecanoyl residues, respectively. The fluorogenic lables and, therefore, the respective fluorescent phospholipids exhibited very different molecular and optical features that made them useful for quite different purposes [4].

If trace amounts of DPH-plasmalogen or -phosphatidylcholine (PC) [5] are incorporated into single bilayer vesicles consisting of unlabeled plasmalogen or PC, emission maxima at 430 nm with similar fluorescence intensities are observed. The measured fluorescence anisotropies for the ether lipid and the diacyl analogue system were also very similar, indicating very similar lipid mobilities in the bilayer. Very profound differences were observed, however for the fluorescence lifetimes of DPH-lipids in bilayers of alkenyacyl and diacyl glycerophosphocholines. The decay times were determined for both systems, using a commercial phase and modulation fluorometer. The samples were excited with sinusoidally modulated laser light. The phase shift and the demodulation of the modulated emission were determined relative to the excitation at different modulation frequencies between 1 and 200 MHz. The measured phase shifts and demodulations could be fitted to unimodal Lorentzian lifetime distributions [6].

Figure 11.3 shows the unimodal Lorentzian lifetime distributions of DPH-plasmalogen and -PC in vesicles of their respective unlabeled analogs. The lifetime centers are rather similar for both systems. However, the distribution

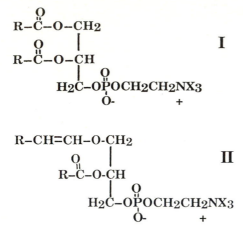

Fig. 11.1. Chemical structures of choline phospholipid subclasses:–I: 1,2 Diacyl-GPC (phosphatidylcholine), II: 1-O-Alkenyl-2-acyl-GPC (choline plasmalogen), GPC = *sn*-glycero-3-phosphocholine, R = aliphatic hydrocarbon chain

widths are much broader for the ether lipid. The same diagram shows, in addition, the effect of cholesterol on the lifetime distributions. Cholesterol is contained in every animal cell and thus, in all membranes that contain plasmalogen [7]. The sterol reduces the lifetime heterogeneity in alkenylacyl and diacyl phospholipid membranes. However, a plasmalogen bilayer containing equimolar amounts of cholesterol is still more heterogeneous than a sterol-free PC membrane.

The lifetime heterogeneity of DPH-lipids basically reflects the heterogeneity of the label environment in the membrane. The observed differences in lifetime heterogeneities between ether and diacyl phospholipid membranes must be a consequence of the differences in the chemical structure of the glycerol region between alkenyl and diacyl glyceorphospholipids. Phospholipid-bound DPH is in close vicinity of the hydrophobic-hydrophilic membrane interface (see Fig. 11.2) and hence is a sensor for dielectric effects in this membrane region. Time-dependent changes in transversal membrane polarity on an appropriate time scale might then cause the observed differences in lifetime distribution widths between plasmalogen and PC. In the meantime, the existence of lifetime distributions could be confirmed independently by pulse fluorometry, showing slightly asymmetric curves for both lipids, that were again much broader for plasmalogen compared with its diacyl analog [8].

Additional evidence that ether and diacyl phospholipids differ with respect to dynamics and polarity in membrane interfaces could be obtained from time-resolved emission spectra of PRODAN, a solvent-sensitive label [9], in vesicle bilayers of alkenylacyl, alkylacyl, or diacyl glycerophosphocholines [10]. It has been well established that PRODAN localizes to the hydrophobic–hydrophilic

Fig. 11.2. Chemical structures of fluorescent choline plasmalogens. Fluorescent acyl residues ▶ in position 2 of glycerol: DPH = diphenylhexatriene propionyl, PYR = pyrene decanoyl, PNA = *trans* parinaroyl, NBD = nitrobenzoxadiazolylaminododecanoyl

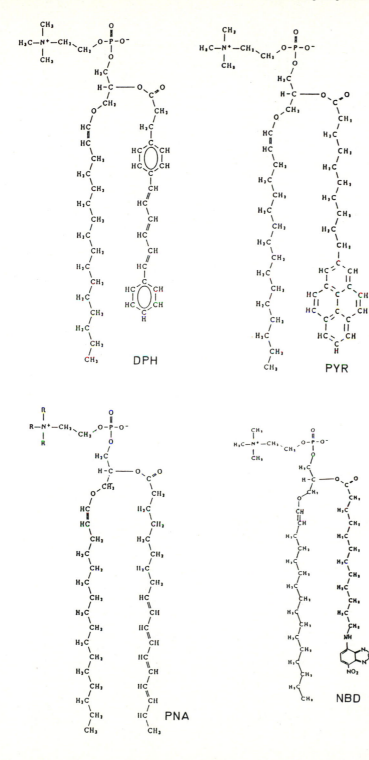

DPH

PYR

PNA

NBD

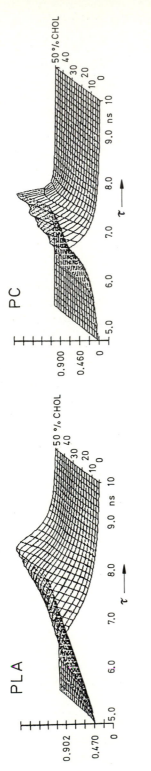

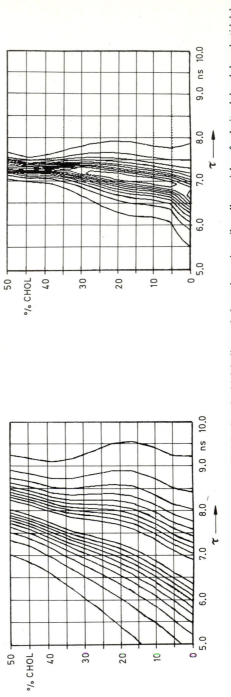

Fig. 11.3. Unimodal Lorentzian lifetime distributions (*top*) of DPH-phosphatidylcholine and plasmalogen in unilamellar vesicles of palmitoyloleoylphosphatidylcholine (PC) and choline plasmalogen (PLA), respectively. Effects of cholesterol on distributional widths (FWHM). The respective contour plots (*bottom*) refer to lifetime distribution widths [ns] at half maximum. The label to phospholipid ratio was 1/300 [mol/mol]

interface of phospholipid bilayers. Thus, it monitors the membrane layer which is chemically different in ether and diacyl phospholipids. PRODAN undergoes solvent relaxation in the respective lipid bilayer systems on a nanosecond time scale. The solvent dipoles, that correspond to the lipid and water dipoles in the membrane interface, undergo reorientation around the excited state of the fluorophor. Solvent reorientation leads to stabilization of the excited state and concomitantly to a time-dependent red shift in the emission.

Inspection of the time-dependent migration of the spectral center of gravity from the unrelaxed to the relaxed excited state showed faster relaxation times for the alkenyl and alkyl ether lipids compared with the diacyl analogue (Fig. 11.4). This faster spectral relaxation must be a consequence of higher molecular flexibility in the interface of ether lipid membranes. The same tendencies could also be observed by deuterium-NMR [11]. Data obtained by this technique provided evidence that the alpha-methylene segments of the acyl chains are more flexible in choline plasmalogen than in diacyl glycerophosphocholine bilayers. Thus, fluorescence and NMR methodes, although on different time scales point to higher segmental mobilities in ether phospholipid membranes.

In addition, the spectral positions of PRODAN in alkenylacyl and alkylacyl glycerophosphocholine are red-shifted relative to the diacyl analog (Fig. 11.4). Accordingly, the ether lipid interface must be more polar than the interface in diacyl phospholipid bilayers. The increased polarity might be a result of more effective water penetration into the glycerol region of ether lipid bilayers, since the simple replacement of an acyl by an enolether group should rather lead to a slight decrease in polarity. In summary, the nanosecond-resolved fluorescence data provide evidence for more effective penetration of water into the hydrophobic–hydrophilic interface of ether lipid bilayers.

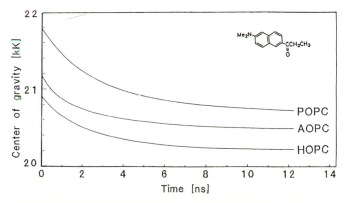

Fig. 11.4. Time-resolved emission spectra of PRODAN in phospholipid bilayers. V_{cg} = center of gravity. POPC = diacyl-GPC, AOPC = alkylacyl-GPC, HOPC = alkenylacyl-GPC. GPC = *sn*-glycero-3-phosphocholine. Insert: Structure of 2-propionyl-6-dimethylamino-naphthalene (PRODAN), (from Ref. 10)

Fig. 11.5. Schematic representation of the spontaneous and protein-catalyzed transfer of phospholipid molecules between bilayer membrane surfaces through the aqueous phase

The high polarity of interfaces in ether phospholipid membranes affects, in turn, other physical properties of this phospholipid subclass, such as intermembrane transfer through the aqueous phase (Fig. 11.5). We investigated the exchange of fluorescent alkenylacyl and diacyl glycerophospholipids between donor vesicle bilayers containing only labeled lipid and unlabeled acceptor systems such as vesicles or biological membranes. The rates of phospholipid transfer were determined from the increase of fluorescence intensity after dilution of labeled lipid into the unlabeled acceptor membrane. The rates for the spontaneous lipid transfer of ether phospholipids were found to be slow but significantly higher than for the transfer of the diacyl analogs. The differences become more evident if the transfer rates are determined for both lipid subclasses in the presence of (PI/PC) specific phospholipid transfer proteins from animal sources such as beef liver or from yeast [12]. These proteins form stoichiometric complexes with choline phospholipids and thus facilitate their transport between membranes through the aqueous phase.

The phenomena discussed so far are rather representative for some general biophysical properties of ether lipids in membranes. More direct information with biological relevance could be obtained when suitable cellular systems became available a few years ago. We used cultured human skin fibroblasts from patients affected with so called peroxisomal diseases such as the cerebro-hepato-renal (CHRS) Zellweger syndrome and rhizomelic Chondrodysplasia Punctata (RCP) [13]. As a consequence of impaired peroxisomal function, the respective cells show drastically reduced plasmalogen levels. The same effect has been observed with the mutant ZR 82 of Chinese hamster ovary (CHO) cells [14]. On the other hand, the plasmalogen-deficient cells synthesize alkenylether lipids if they are cultured in the presence of suitable precursors that circumvent the ether lipid biosynthesis block in the peroxisomes. Hexadecyl glycerol containing a preformed glycerol ether bond is such a suitable substrate [15] that increases the plasmalogen content of CHRS, RCP, and ZR 82 cells to almost normal levels. Therefore, the cells enriched in ether lipids by supplementation with alkyl glycerol provide a very nice reference system for the studying effects of plasmalogens on the properties of biological membranes.

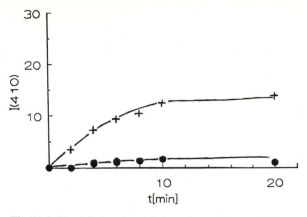

Fig. 11.6. Uptake of parinaroyl plasmalogen from vesicles into cultured human skin fibroblasts as determined from the increase of fluorescence intensity at 410 nm (I(410)). Cell monolayers (22 μg phospholipid) grown to confluency on glass cover slips were incubated with the fluorogenic vesicles (9 nmol phospholipid) in a cuvette at 37°C. + = plasmalogen-deficient CHRS cells (see text), ● = plasmalogen-containing CHRS cells (see text)

First of all, it was interesting to find out whether plasmalogen deficiency of the cell membranes would increase the tendency of the cells to compensate this deficiency by preferred uptake of exogenously added ether phospholipid. For these experiments, cells were grown on glass cover slips and incubated with vesicles containing the fluorescent phospholipid. The rates of lipid incorporation into cell monolayers were determined from the continuous increase in fluorescence intensity. By separate experiments we were able to confirm that uptake of fluorescent phospholipid into the plasma membrane was exclusively due to transfer from vesicle to cell surfaces.

Figure 11.6 shows the transfer of parinaroyl plasmalogen and PC into CHRS cells. The plasmalogen-deficient cells show the highest uptake rates for the ether lipid [16]. In contrast, plasmalogen-enriched CHRS cells or control cells from healthy donors incorporate less plasmalogen. Furthermore, only ether lipid uptake into the deficient cells is increased. Transfer of diacyl glycerophosphocholine into cell plasma membranes is insignificant with all cell strains under investigation. The same effect, regulation of plasmalogen uptake by the plasmalogen content of the cells, was also observed for RCP cells and the ether lipid deficient hamster ovary cells. Therefore, we may conclude that the observed differences in transfer rates are effects due to the differences in the ether lipid content of the various cell types. In addition, plasmalogen uptake was always favoured over diacyl phospholipid uptake irrespective of the fluorescent acyl chain in position 2 of glycerol. Parinaroyl, pyrenedecanoyl, DPH propionyl and NBD dodecanoyl alkenylacyl and diacyl lipid pairs always showed the same differences.

Fluorescence microscopy provided evidence that the fluorescent ether lipids are transferred in a first step from vesicles to the plasma membrane before they

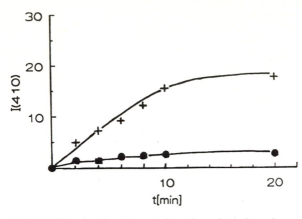

Fig. 11.7. Transfer of parinaroyl plasmalogen into plasmalogen-containing Chinese hamstrer ovary (CHO) cells (controls) as determined from the increase of fluorescence intensity at 410 nm (I(410)) (for concentrations see legend to Fig. 11.6). (+) in the presence and (●) in the absence of supernatant from plasmalogen-deficient CHO cells (see text)

eventually entered the cell interior followed by biochemical degradation. When monolayers of plasmalogen-deficient CHO cells were incubated with NBD-plasmalogen for a few seconds at 37°C, the cell surface is stained, whereas intracellular membranes become visible only after several minutes.

We addressed ourselves to the question as to the molecular origin of facilitated plasmalogen transport from vesicles into cells. A first hint came from experiments with aqueous buffer media that were incubated with monolayers of plasmalogen-deficient cells grown on glass cover slips. Cultured cells with normal plasmalogen levels do not incorporate exogenous ether lipid under normal conditions. However, they take up fluorescent ether lipid if supernatant from plasmalogen-deficient RCP cells is added (Fig. 16.7). In the absence of such a catalyst no transfer is observed. Evidently, a factor is released from the deficient cells into the aqueous medium, which is responsible for this effect.

The plasmalogen transfer factor can be isolated from any of the available plasmalogen deficient cell strains of human (CHRS, RCP) or animal (CHO) origin. Its presence on and release from the cell surface is strictly dependent on the plasmalogen content of the cells. In culture media of cells that were enriched with plasmalogen or media of control cells with normal plasmalogen levels, such transfer activity could not be detected. A series of biochemical characterization procedures provided evidence for a protein-like nature of the transfer factor.

According to results obtained from kinetic experiments [17] with fluorescent ether lipids, this putative plasmalogen transfer protein acts as a shuttle for the transport of plasmalogens through the aqueous phase. It probably forms a 1/1 complex with the ether lipid and thus obeys a similar mechanism as already postulated for the PI/PC specific cytosolic phospholipid transfer proteins [18]. However, the plasmalogen transfer protein is a cell surface protein. In addition, it is highly selective for enolether bonds and not the polar head groups in choline and ethanolamine phospholipid molecules. This is an absolute novelty in the field of lipid-protein interactions.

References:

1. Snyder F (1972) Ether lipids: chemistry and biology. Academic Press, New York
2. Mangold HK, Paltauf F (1983) Ether lipids: biochemical and biomedical aspects. Academic Press, New York
3. Hermetter A (1988) Comm Mol Cell Biophys 5: 133
4. Hermetter A (1990) Appl Fluor Technol 2: 1
5. Morgan CG, Thomas EW, Moras TS, Yanni YP (1982) Biochim Biophys Acta 735: 418
6. Hermetter A, Kalb E, Paltauf F (1989) Biophys J 56: 1245
7. Yeagle P (1985) Biochim Biophys Acta 822: 267
8. Kungl AJ, Hermetter A, Paltauf F, Kauffmann HF, Prenner E (1991) In: Hester RE, Girling RB (eds) Spectroscopy of biological molecules. The Royal Society of Chemistry, Cambridge
9. Weber G, Farris FJ (1979) Biochemistry 18: 3075
10. Sommer A, Paltauf F, Hermetter A (1990) Biochemistry 29: 1134
11. Malthaner M, Hermetter A, Paltauf F, Seelig J (1987) Biochim Biophys Acta 900: 191
12. Szolderits G, Daum G, Paltauf F, Hermetter A (1991) Biochim Biophys Acta 1063: 197
13. Schutgens RBH, Heymans HSA, Wanders RJA, v.d. Bosch H, Tager JM (1986) Eur J Pediatr 144: 430
14. Zoeller RA, Raetz RH (1986) Proc Natl Acad Sci USA 83: 5170
15. Paltauf F, Hermetter A (1989) J Clin Chem Clin Biochem 27: 287
16. Loidl J, Schwabe G, Paschke E, Paltauf F, Hermetter A (1990) Biochim Biophys Acta 1049: 75
17. Yoshimura T, Welti R, Helmkamp JGM (1988) Arch Biochem Biophys 266: 299
18. Khader JC, Douady D, Mazliak P (1982) In: Hawthorne JN, Ansell GB, Form and function of phospholipids. Elsevier, Amsterdam, pp 279–311.

12. Pyrene-Labelled Lipids as Fluorescent Probes in Studies on Biomembranes and Membrane Models

Paavo KJ Kinnunen*, Anu Koiv and Pekka Mustonen

Lipid Research Laboratory, Department of Medical Chemistry, University of Helsinki, Siltavuorenpenger 10, SF-00170 Helsinki, Finland

1 Introduction

The immense progress in the understanding of the functions of proteins and nucleic acids in living cells is still contrasted by the limited comprehension of the significance of lipids and their structural diversity in the self-assembly and functionalization of biological membranes. Nevertheless, taking into account the costly maintenance of the specific tissue, cell, and cell organelle lipid compositions and their alterations it is obvious that the individual lipid classes do serve more important roles than what is at present commonly accepted [1].

The use of physicochemical techniques has been fundamental in obtaining data and tools for the analysis and description of the complex properties of lipids. A vast array of methods is currently utilized; however, one of the most popular is fluorescence spectroscopy. This is due to the relatively easy instrumentation as well as the commercial availability of lipid derivatives covalently labelled with various fluorophores. Like ESR this approach requires the introduction of a bulky group into the membrane. The resulting perturbation by fluorescent lipid probes is a clearly unavoidable disadvantage. Yet, under carefully controlled experimental conditions they do allow access to the observation of the various membrane properties.

Amongst the different fluorescent moieties covalently linked with various lipid structures pyrene has gained increasing popularity and has been so far applied to study a large number of properties of biomembranes and membrane models (Table 12.1). Some pyrene-phospholipids exhibit thermotropic phase transitions and have been analyzed in considerable detail [2, 3]. Because of the limited space available this brief review focuses on research and topics reflecting some of the interests of our own laboratory.

2 Basic Photophysics of Pyrene and Its Lipid Derivatives

In addition to its normal fluorescence features pyrene can perhaps be considered as a paradigm of excimer (from *exci*ted *di*mer) forming aromatic hydrocarbons.

---

* Senior investigator of the Finnish State Medical Research Council

Table 12.1. Examples of applications of pyrene-labelled lipids in studies on biomembranes and membrane models

---

Characterization of liposomes and lipid dynamics:
– Phospholipid phase transitions [4–7]
– Acyl chain interdigitation [8]
– Acyl chain alignment [9, 10]
– Phase separation [11–14]
– Lipid distribution [5, 15–17]
– Fluidity [4, 18]
– Lateral diffusion [4]
– Lipid flip-flop [19]
– Liposome surface pressure [20, 21]
Characterization of lipid monolayers [11, 22–24]
  & Langmuir–Blodgett films [25–27]
Membrane fusion [28]
Lipid transfer between vesicles & plasma lipoproteins:
– spontaneous [29–33]
– protein mediated [34–36]
Integral membrane protein–lipid interactions [37–39]
Membrane attachment of:
– drugs [40, 41]
– peptides [4, 12]
– peripheral proteins [13, 14]
– polysomes [42]
Action of lipolytic enzymes:
– phospholipase A_2 & C [9, 43–45]
– sphingomyelinase [46]
– triacylglycerol hydrolases [47, 48]
– cholesterylester hydrolase [49]
Fatty acid and lipid uptake by cells [50–55]
Lipid peroxidation [56]

This property of pyrene was discovered in 1954 by Förster and Kasper [57, 58]. In a simplified schema excimer formation can be described as follows:

$$P + hv \rightarrow P^* \rightarrow P + hv \approx 375/400, \tag{1}$$

$$P + P^* \rightarrow [P_2]^* \rightarrow P + P + hv \approx 480. \tag{2}$$

In brief, monomeric pyrene relaxes back to the ground state by emitting fluorescence with maximum around 375 or 400 nm, depending on solvent polarity [59]. Increasing the concentration of pyrene, however, results in radiative relaxation through a pathway emitting photons as a broad and structureless band centered at ≈ 480 nm. This shift to lower energies occurs due to the formation of a complex consisting of two pyrenes in close proximity (≈ 3.35 Å) and attracted in a potential well. The latter, in turn, is due to the formation of a short-lived, occupied, and bonding π^* orbital. After relaxation two monomeric ground state pyrenes are released. Detailed discussion on excimer photophysics has been written by Birks [60].

Depending on the applications in question the long lifetime of pyrene may represent an advantage. In addition it has a high quantum yield. Pyrene is, however, poorly suited for fluorescence polarization studies. Most applications of pyrenelipids have exploited the measurement of I_e/I_m while more sophisticated features of the fluorescence properties of these probes still remain largely unexplored, evidently because of the high cost of instrumentation required for the acquisition and analysis of fast, time-resolved fluorescence spectra. The photophysics of pyrene is of course of interest as such. Notably, the utilization of pyrenelipids offers novel means to study not only the properties of lipid membranes but also allows to explore the kinetics and mechanisms of excimer formation in more detail [61–63].

Pyrenelipids have also been used in fluorescence energy transfer studies taking advantage of the spectral overlap between tryptophan emission and pyrene excitation [39]. In this elegant study by Verbist et al. a positive correlation was found between the number of negative charges in the head groups of different pyrene-containing phospholipids and the degree of their association with the erythrocyte ($Ca^{2+} + Mg^{2+}$) ATPase. Resonance energy transfer takes place also between pyrene (donor) and the heme moiety of a peripheral membrane protein, cytochrome c (acceptor), thus resulting in fluorescence quenching [14]. The sensitivity of the fine structure of pyrene monomer emission to probe environment has been utilized to observe phospholipid acyl chain interdigitation [8]. One drawback of pyrene is its quenching by oxygen. However, this property can be made use of in monitoring the content of oxygen in membranes [64].

Due to its lack of polarity pyrene readily accommodates in the hydrocarbon region of the membrane and the initial studies did employ pyrene as such. However, the potential of more specific structural lipid analogs was soon

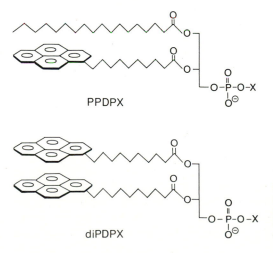

PPDPX

diPDPX

Fig. 12.1. Chemical structures of the inter- and intramolecular excimer forming phospholipid probes PPDPX and bisPDPX, respectively. X may be serine, choline, glycerol etc.

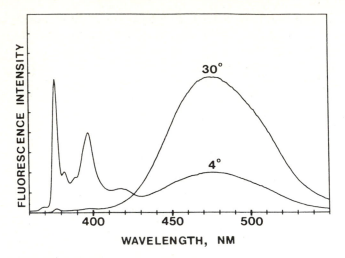

Fig. 12.2. Normalized (at peak intensities) fluorescence emission spectra for PPDPC liposomes in water, measured at 4 and 30 °C

realized and, consequently, pyrene-labelled lipids such as 10-(pyren-1-yl)-decanoic acid (PDA) and 1-palmitoyl-2-[10-(pyren-1-yl)]-decanoyl-sn-glycero-3-phosphocholine (PPDPC) were synthesized [4]. Since then all the major lipid classes, cholesterol and its fatty acid esters, tri- and diacylglycerols, fatty acids with different chain lengths, ceramides, and different phosphoglycerides have been prepared. In addition to the intermolecular excimer forming analogs with a single pyrene moiety, also lipid probes containing two fluorophores and thus capable of intramolecular excimer formation have been used. First excimeric pyrenephospholipid was 1,2-bis-[10-(pyren-1-yl)]-decanoyl-sn-glycero-3-phosphocholine (bisPDPC, [6]). Examples of these two types of pyrenelipids are illustrated in Fig. 12.1. Emission spectra for PPDPC liposomes below (at 4°C) and above (at 30°C) their phase transition are shown in Fig. 12.2. The distinct features of monomer and excimer emission are clearly evident.

3 Studies on Lipid Dynamics

One of the interesting features of different lipids is their ability to undergo phase transitions which cause drastic changes in the membrane properties, lipid conformation, conformational dynamics, lateral motion, as well as in the two- and three-dimensional organization of such lipid assemblies [65]. While thermotropic transitions have been studied in depth it is likely that isothermally induced transitions bear more physiological relevance. Factors inducing the latter type phase changes include proteins, ions, and pH. Notably, cell membranes undergo dramatic alterations in their morphology following for instance stimulation by growth factors [66, 67]. Changes in [plasma membrane area]/ [intracellular volume] provide attractive means for triggering lipid phase transitions in cells [68].

Pyrene lipids have been applied by several groups to study phospholipid phase transitions [4–7]. Using PPDPC (which forms intermolecular excimer only) the increased rate of lateral diffusion in the liquid crystalline membrane is evident as an increased rate of excimer formation. Incorporation of the excimeric probe bisPDPC at very low concentrations into a matrix of 1,2-dipalmitoyl-*sn*-glycero-3-phosphocholine (DPPC) allows to observe intramolecular excimer only [6]. Interestingly, under these conditions an increase in I_e/I_m occurs accompanying the transition of DPPC at 41 °C. This observation was suggested to be due to intercalation of the acyl chains of those lipids in the other leaflet of the phospholipid bilayer between the pyrene moieties of bisPDPC when the DPPC matrix is maintained at $T < T_m$, i.e. in the gel state. Another possibility is a change in the lipid conformation altering the vertical alignment of the acyl chains similarly to that suggested for the acidic excimeric phospholipid 1,2-bis-[10-(pyren-1-yl)]-butanoyl-*sn*-glycero-3-phospho-*rac*-glycerol (bisPBPG) and induced by increasing the ionic strength of the aqueous phase [9, 10].

3.1 Phospholipid Phase Separation

In proper lipid mixtures complete mixing may not take place but instead phase separation occurs. Phase separations frequently accompany phospholipid phase transitions. Isothermal phase separations can be induced by several factors such as membrane proteins, ions, pH, and membrane potential. Such phase separation processes can be directly visualized by fluorescence microscopy of cells [69], liposomes [70], and lipid monolayers [11, 71]. As an example of the latter, Ca^{2+} induced phase separation in a monolayer of DPPC/PPHPA (80:20, molar ratio) and residing on an argon/water interface is shown in Fig. 12.3. The physiological significance of phospholipid phase separation is still unknown.

Cytochrome c (cyt c) is an integral component of the mitochondrial respiratory chain transfering electrons to cytochrome c oxidase. At the moment cyt c is probably the best characterized peripheral membrane protein. In our studies on the electrostatic attachment of cyt c to liposomes we took advantage of the spectral overlap between pyrene fluorescence and the cyt c heme absorption [14]. The pyrene lipid contained the fluorophore embedded in the membrane hydrocarbon region. Therefore, very little if any perturbation of the membrane surface attachment of cyt c to acidic phospholipids is anticipated. To summarize our findings we could show that the binding of this peripheral membrane protein could be regulated by phase transition induced phase separation. In brief, we used DPPC liposomes containing 5 mol% of phosphatidic acid (PA). When the liposomes are maintained at $T > T_m$ the negatively charged PA is dissolved in the liquid crystalline DPPC. At this low content of PA the surface charge density is too low to provide a binding site for cyt c carrying a net

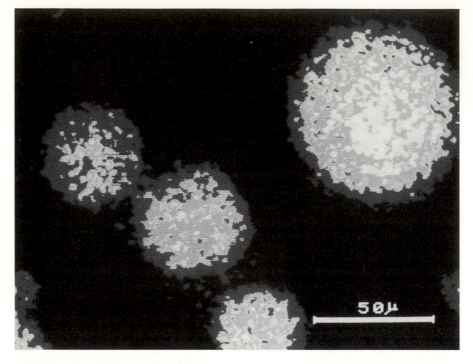

Fig. 12.3. Fluorescence image of a monolayer of DPPC/1-palmitoyl-2-[10-(pyren-1-yl)]-hexanoyl-*sn*-glycero-3-phosphate, PPHPA (80:20, molar ratio) on an argon/water interface at a surface pressure of 10 mNm^{-1}. Content of Ca^{2+} in the subphase = 1 mM

positive charge. However, when temperature is lowered a phase separation occurs as the fluid PA is excluded from the crystallizing DPPC membrane. As a consequence, negatively charged membrane domains are formed which now attach cyt c to the liposome surface [14]. To conclude, phase separation in membranes leading to patches of negatively charged lipids could be utilized by cells to regulate their electrostatic contacts with proteins.

3.2 Formation of Lipid Superlattices

An unexpected finding of our studies on the dependency of I_e/I_m on the content of PPDPC in liquid crystalline matrices of either DPPC or eggPC was that increasing the amount of the pyrenelipid resulted in the formation of well defined regularities in the slopes of I_e/I_m vs [PPDPC] (Ref. 5). These data were substantiated using Langmuir–Blodgett technique, i.e. transferring PPDPC/DPPC monolayers from an air/water interface onto quartz slides [17]. The observed discontinuities could be reproduced with reasonable accuracy by simple geometrical model based on a hexagonal distribution of the pyrenede-

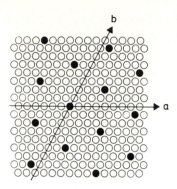

Fig. 12.4. Illustration of a superlattice of PDA chains (*filled dots*) of PPDPC in a hexagonal acyl chain lattice. The superlattice coordinates correspond to 10.5 mol% PPDPC in a matrix of a natural phospholipid

canoyl chains of PPDPC in a hexagonal acyl chain lattice [72]. The superlattice spacing is dependent on the concentration of PPDPC. An example is illustrated in Fig. 12.4. The driving force is likely to be steric elastic strain due to the bulky pyrene moiety which exerts pressure in its vicinity and thus causes a repulsive potential between PPDPC molecules representing substitutional impurities.

3.3 Estimation of Liposome Equilibrium Lateral Surface Pressure

Equilibrium lateral surface pressure is an important parameter influencing membrane properties. For instance, the membrane penetration of peptides and the action of lipolytic enzymes is critically dependent on the lipid packing density (e.g. [73]). Unfortunately, surface pressure in biomembranes and liposomes is very difficult to address experimentally. In an effort to develop lateral packing pressure sensitive probes we utilized the spectral sensitivity of pyrene absorption to a vicinal π electron system [21]. For this purpose we synthesized a fluorescent probe, 1-palmitoyl-2-[10-(pyren-1-yl)-10-ketodecanoyl]-*sn*-glycero-3-phosphocholine (k-PPDPC). Two endothermic transitions were observed by differential scanning calorimetry for neat k-PPDPC liposomes at 13.7 and 23.8 °C. Coinciding with the second calorimetric transition the ratio of the absorption bands at 289 and 356 nm changed from 2.15 to 1.60. Thereafter we measured reflectance spectra for k-PPDPC monolayers on water so as to determine the intensities of these bands as a function of surface pressure. Comparison of the data recorded for liposomes and monolayers allowed us to estimate a surface pressure decrease in k-PPDPC liposomes from ≈ 39 to 17 mN/m accompanying the thermotropic transition at 23 °C.

In a related effort we measured compression isotherms for monolayers of 1-palmitoyl-2-[10-(pyren-1-yl)]-hexanoyl-*sn*-glycero-3-phosphocholine (PPHPC), and the corresponding -methanol, -serine, -glycerol, and -ethanolamine derivatives on an air/water interface [20]. Thereafter we recorded I_e/I_m for liposomes of the same phospholipids. We then sought for a surface pressure value giving best correlation between (i) mean molecular area and (ii) the measured I_e/I_m values for liposomes. This procedure indicates an estimated surface pressure of ≈ 12 mN/m in liposomes consisting of these phospholipids. Notably, we assumed the surface pressures to be identical irrespective of the head group. Additionally, we assumed I_e/I_m to be linearly dependent on r^{-2} ($r =$ the average distance between the pyrene moieties). Accordingly, I_e should be linearly dependent on r^{-1}. This is in accordance with data on pyrenelipid containing Langmuir–Blodgett films [17] and monolayers [11].

4 Peripheral Ligand–Membrane Interactions

The regulation of the spatio-temporal architecture of cell membranes has been suggested to be of utmost physiological significance and apparently regulated by factors such as peripheral interactions of soluble ligands (e.g. proteins, ions, and drugs) on lipid membrane surfaces [1]. As examples, more detailed account on the membrane attachment of cytochrome c, and polysomes is given below.

4.1 Formation of Superlattices of Membrane-Bound Cytochrome c

Regulation of the membrane binding of cyt c by phospholipid phase separation was already discussed above. In the following data indicating the formation of hexagonal superlattices of cyt c on liposome surface is briefly reviewed.

In this study we took advantage of the measurement of pyrene fluorescence lifetime. More specifically, due to the membrane attachment of cyt c energy transfer from pyrene to the heme group takes place. This quenching of pyrene emission leads not only to the decrease in pyrene emission intensity but also to the decrease of fluorescence lifetime τ. The latter parameter is highly sensitive to the physical state of the immediate environment of the fluorophore. Additionally, it is not disturbed by trivial absorption of the emitted light by the cyt c heme. Measurement of τ reflects the probe environment beyond the quenching radius of cyt c. Accordingly, it provides a measure of the long range membrane perturbation by surface attached cyt c. We used 1-palmitoyl-2[10-(pyren-1-yl)]-decanoyl-sn-glycero-3-phosphate (PPDPA) embedded in a fluid matrix of either eggPC or DPPC. Concentration of PA was chosen to be high enough to provide sufficient negative surface charge density to attach cyt c from the solution to the liposome surface. Somewhat unexpectedly τ vs [cyt c] plots revealed stepwise decrements, such as those illustrated in Fig. 12.5. The only reasonable explanation to fit these data is the formation of regular superlattices of membrane bound cyt c. The lattice constants could be reproduced with a simple geometrical model. The basic features of this model are illustrated in Fig. 12.6. For more detailed discussion the reader is referred to the original publication [14].

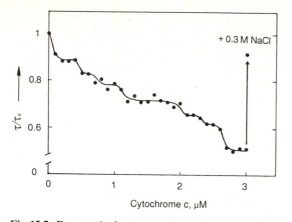

Fig. 12.5. Decrease in the average relative fluorescence lifetime of 1 mol% 1-palmitoyl-2-[10-(pyren-1-yl)]-decanoyl-*sn*-glycero-3-phosphate (PPDPA) as a function of cyt c concentration. Total lipid was 200 µM and the content of eggPA in the small unilamellar liposomes was 19 mol%, the rest being eggPC. The initial fluorescence lifetime in the absence of cyt c was $127 + 3$ ns. The *upward arrow* on the right illustrates the effect of displacing cyt from the liposome surface by 0.3 *M* NaCl

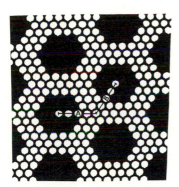

Fig. 12.6. Hexagonal superlattice of cyt c binding sites in a liquid crystalline liposome membrane. Each binding site consists of 19 acyl chains, corresponding to 9.5 phospholipids

4.2 Binding of Polysomes to Rough Endoplasmic Reticulum Membrane

Synthesis of secreted as well as integral membrane proteins takes place perhaps exclusively in the rough endoplasmic reticulum, RER, which has derived its name because of the granular electromicroscopic appearance due to the membrane bound polyribosomes, polysomes. These can be removed by KCl/puromycin treatment to yield degranulated RER (dRER). There are indications of the participation of lipids in the regulation of the translation process in RER (for a brief review, see Ref. 42). As a first step of our effort to elucidate the possible role of lipids in protein synthesis in RER, we developed a technique, differential fluorescence quenching (DEQ) to monitor changes in the distribution and dynamics of RER lipids, occurring as a consequence of the membrane attachment of polysomes to dRER labelled with pyrenedecanoic acid [42].

Two sets of data were collected. First, changes in the I_e/I_m of PDA containing dRER were monitored upon the addition of native polysomes to the membrane preparation. Obviously, these changes in pyrene fluorescence arise from perturbation produced by the membrane bound polysomes and reflect changes in the entire membrane. The second set of data was collected using trinitrophenylated polysomes (TNP-polysomes). Because of the spectral overlap between pyrene emission and TNP absorption [44] this moiety allows energy transfer to be used to assess the binding of TNP-polysomes to PDA containing dRER. The latter fluorescence signals arise in the membrane beyond the quenching radii of the polysome-linked TNP. An attempt to clarify the prevailing situation is provided in Fig. 12.7. Accordingly, simple arithmetics allows to obtain the I_e/I_m values for the membrane vicinal to the membrane bound polysome:

$$I_{site} = I_0 + (I_{native} - I_{TNP}), \tag{3}$$

where:

I_{site} the pyrene fluorescence intensity (monomer or excimer) in the membrane domain accommodating the polysome on its surface

I_0 the initial fluorescence intensity recorded for dRER prior to the addition of polysomes (native or TNP-labelled)

I_{native} fluorescence intensity measured after the addition of native polysomes, and

I_{TNP} fluorescence intensity measured after the addition of TNP-labelled polysomes.

In order to estimate to what extent the observed changes in I_e/I_m in the two membrane domains reflect (i) changes in the diffusion rate and (ii) changes in the lateral distribution of PDA, we defined parameter Q as follows:

$$Q = I_m + bI_e \tag{4}$$

If the number of quanta emitted is assumed to be independent from the I_e/I_m when no quenching occurs, this parameter is related to quantum yield and should be constant.

The value for the constant b (dependent on the instrument when uncorrected spectra are used) was then obtained by measuring I_m and I_e for a

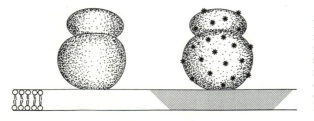

Fig. 12.7. Schematic view of a native (*left*) and a TNP-labelled polysome (*right*) attached on the surface of endoplasmic reticular membrane. *Shaded membrane area* corresponds to the quenching radii of the polysome-linked TNPs

polysome free membrane (Q_1) and after the addition of native polysomes (Q_2). These values for Q should be equal and we thus obtain the numerical value for b, in our case 0.33. We can then plot $Q = I_m + 0.33\, I_e$ as a function of native and TNP-labelled polysomes. As expected, Q vs [native polysomes] remains essentially constant, whereas Q vs [TNP-polysomes] revealed biphasic decrease. In brief, when using pyrene labelled lipids the above procedure, nominated by us as *differential fluorescence quenching*, allows one to estimate qualitative changes in membrane lipid dynamics, yielding information both on the diffusibility of the probe as well as on its distribution (lateral and transmembrane) in the membrane. This technique should be applicable to membrane proteins in general, assuming that labelling of the protein by a quenching group does not perturb the system under study too seriously.

5 Future Perspectives

The next decade will certainly provide us with further important clues regarding the roles of lipids in the structure and function of biological membranes and subsequently leading into similar breakthroughs in our understanding on the cell properties as we have witnessed after the elucidation of the genetic code. Yet, the field of lipids still warrants major efforts. We may anticipate fluorescence spectroscopy techniques to continue to increase in popularity in these studies. Learning the principles of sorting different lipid structures in cells, together with improved fluorophores and instrumentation, such as quantitative, energy- and time-resolved fluorescence microscopy, combined with thorough understanding of the probe photophysics, will certainly represent some of the next developments.

6 References

1. Kinnunen PKJ (1991) Chem Phys Lipids 57: 375
2. Lotta TI, Laakkonen LJ, Virtanen JA, Kinnunen PKJ (1988) Chem Phys Lipids 46: 1
3. Lotta TI, Virtanen J, Kinnunen PKJ (1988) Chem Phys Lipids 46: 13
4. Galla HJ, Hartmann W (1980) Chem Phys Lipids 27: 199
5. Somerharju PJ, Virtanen JA, Eklund KK, Vainio P, Kinnunen PKJ (1985) Biochemistry 24: 2773
6. Sunamoto J, Kondo H, Nomura T, Okamoto H (1980) J Am Chem Soc 102: 1146
7. Chen S-Y, Cheng KH, Ortalano DM (1990) Chem Phys Lipids 53: 321
8. Komatsu H, Rowe ES (1991) Biochemistry 30: 2463
9. Thuren T, Vainio P, Virtanen JA, Somerharju P, Blomqvist K, Kinnunen PKJ (1984) Biochemistry 23: 5129
10. Kinnunen PKJ, Thurén T, Vainio P, Virtanen JA (1985) In: Degiorgio V, Conti M (eds) Physics of Amphiphiles: Micelles, Vesicles and Microemulsions, Elsevier North-Holland, p 687
11. Eklund KK, Vuorinen J, Mikkola J, Virtanen JA, Kinnunen PKJ (1988) Biochemistry 27: 3433
12. Jones ME, Lentz BR (1986) Biochemistry 25: 567
13. Wiener JR, Pal R, Barenholz Y, Wagner RR (1985) Biochemistry 24: 7651
14. Mustonen P, Virtanen J, Somerharju PJ, Kinnunen PKJ (1987) Biochemistry 26: 2991
15. Hresko RC, Sugar IP, Barenholz Y, Thompson TE (1986) Biochemistry 25: 3813

16. Hresko RC, Sugar IP, Barenholz Y, Thompson TE (1987) Biophys J 51: 725
17. Kinnunen PKJ, Tulkki AP, Lemmetyinen H, Paakkola J, Virtanen JA (1987) Chem Phys Letters 136: 539
18. Dudeja PK, Harig JM, Ramaswamy K, Brasitus TA (1989) J Am Physiol 257: G809
19. Homan R, Pownall HJ (1988) Biochim Biophys Acta 938: 155
20. Thuren T, Virtanen JA, Kinnunen PKJ (1986) Chem Phys Lipids 41: 329
21. Konttila R, Salonen I, Virtanen JA, Kinnunen PKJ (1988) Biochemistry 27: 7443
22. Loughran T, Hatlee MD, Patterson LK, Kozak JJ (1980) J Chem Phys 72: 5791
23. Subramanian R, Patterson LK (1985) J Am Chem Soc 107: 5821
24. Caruso F, Grieser F, Murphy A, Thistlethwaite P, Urquhart R, Almgren M, Wistus E (1991) J Am Chem Soc 113: 4838
25. Kinnunen PKJ, Virtanen JA, Tulkki AP, Ahuja RC, Möbius D (1985) Thin Solid Films 132: 193
26. Yamazaki T, Tamai N, Yamazaki I (1986) Chem Phys Letters 124: 326
27. Yliperttula M, Lemmetyinen H, Mikkola J, Virtanen JA, Kinnunen PKJ (1988) Chem Phys Lett 152: 61
28. Eklund KK (1990) Chem Phys Lipids 52: 149
29. Doody MC, Pownall HJ, Kao YJ, Smith LC (1980) Biochemistry 19: 108
30. Roseman MA, Thompson TE (1980) Biochemistry 19: 439
31. Correa-Freire M, Barenholz Y, Thompson TE (1982) Biochemistry 21: 1244
32. Massey JB, Gotto AM, Pownall HJ (1982) J Biol Chem 257: 5444
33. Massey JB, Hickson D, She HS, Sparrow JT, Via DP, Gotto AM, Pownall HJ (1984) Biochim Biophys Acta 794: 274
34. Via DP, Massey JB, Vignale S, Kundu SK, Marcus DM, Pownall HJ, Gotto AM (1985) Biochim Biophys Acta 837: 27
35. Massey JB, Hickson-Bick DP, Via AM, Gotto AM, Pownall HJ (1985) Biochim Biophys Acta 835: 124
36. van-Paridon PA, Gadella TW, Somerharju PJ, Wirtz KW (1988) Biochemistry 27: 6208
37. Freire E, Markello T, Rigell C, Holloway PW (1983) Biochemistry 22: 1675
38. Jones OT, Lee AG (1985) Biochemistry 24: 2195
39. Verbist J, Gadella TWJ, Raeymaekers L, Wuytack F, Wirtz KWA, Casteels R (1991) Biochim Biophys Acta 1063: 1
40. Michelangeli F, Robson MJ, East JM, Lee AG (1990) Biochim Biophys Acta 1028: 49
41. Mustonen P, Kinnunen PKJ (1991) J Biol Chem 266: 6302
42. Kaihovaara P, Raulo E, Kinnunen PKJ (1991) Biochemistry 30: 8380
43. Thuren T, Virtanen JA, Kinnunen PKJ (1986) J Membr Biol 92: 1
44. Thuren T, Virtanen JA, Somerharju PJ, Kinnunen PKJ (1988) Anal Biochem 170: 248
45. Thuren & Kinnunen (1991) Chem Phys Lipids 59: 69
46. Klar R, Levade T, Gatt S (1988) Clin Chim Acta 176: 259
47. Salvayre R, Negre A, Radom J, Douste-Blazy L (1986) Clin Chem 32: 1532
48. Negre A, Salvayre R, Dousset N, Rogalle P, Dang QQ, Douste-Blazy L (1988) Biochim Biophys Acta 963: 340
49. Joutti A, Kotama L, Virtanen JA, Kinnunen PKJ (1985) Chem Phys Lipids 36: 335
50. Levade T, Gatt S (1987) Biochim Biophys Acta 918: 250
51. Gatt S, Nahas N, Fibach E (1988) Biochem J 253: 377
52. Pownall HJ, Smith LC (1989) Chem Phys Lipids 50: 191
53. Fibach E, Rachmilevitz EA, Gatt S (1989) Leuk Res 13: 1099
54. Radom J, Salvayre R, Levade T, Douste-Blazy L (1990) Biochem J 269: 107
55. Naylor BL, Picardo M, Homan R, Pownall HJ (1991) Chem Phys Lipids 58: 111
56. Viani P, Cervato G, Cestaro B (1991) Biochim Biophys Acta 1064: 24
57. Förster Th, Kasper K (1955) Z Elektrochem 59: 976
58. Förster Th (1969) Angew Chem 81: 364
59. Dong DC, Winnik MA (1984) Can J Chem 62: 2560
60. Birks JB (1975) Rep Prog Phys 38: 903
61. Lemmetyinen H, Yliperttula M, Mikkola J, Virtanen JA, Kinnunen PKJ (1989) J Phys Chem 93: 7170
62. Barenholz Y, Cohen, Korenstein R, Ottolenghi M (1991) Biophys J 59: 110
63. Sugar IP, Zeng J, Chong PL-G (1991) J Phys Chem 95: 7524
64. O'Loughlin MA, Whillans DW, Hunt JW (1980) Radiation Res 84: 477
65. Kinnunen PKJ, Laggner P (eds) (1991) Phospholipid phase transitions, Chem Phys Lipids 57

66. Goshima K, Masuda A, Owaribe K (1984) J Cell Biol 98: 801
67. Mellström K, Heldin C-H, Westermark B (1988) Exp Cell Res 177: 347
68. Needham D, Evans E (1988) Biochemistry 27: 8261
69. Yechiel E, Edidin M (1987) J Cell Biol 105: 755
70. Haverstick DM, Glaser M (1988) J Cell Biol 106: 1885
71. Weiss RM (1991) Chem Phys Lipids 57: 227
72. Virtanen JA, Somerharju P, Kinnunen PKJ (1988) J Mol Electronics 4: 233
73. Vainio P, Virtanen JA, Kinnunen PKJ, Voyta JC, Smith LC, Gotto AM, Sparrow JT, Pattus F, Verger R (1983) Biochemistry 22: 2270

13. Optical Detection of Intracellular Ion Concentrations

Jan Slavík

Institute of Physiology, Czechoslovak Academy of Sciences, Videňská 1083, CS-142 20, Prague 4-Krč, Czechoslovakia

1 Introduction

Fluorescence techniques give better results in biological and medical applications than do absorption techniques because of their higher sensitivity (the absorption signal is related to the 100% incident light intensity, while in fluorescence small signals are detected against zero background = darkness) and easier separation of the dye-related signal from the background.

Dyes are employed either as *labels*, which simply mark the position of the labeled compound inside the cell, or as *probes*, which give information about what goes on in their molecular neighborhood. Probes act as tiny molecular reporters – "spies" which, encoded in the fluorescence signal, pass on information about their molecular neighbourhood such as polarity, viscosity, ion composition, presence of other chromophores, electrical field, etc. (Fig. 13.1).

The greatest advantage of optical techniques is the possibility to view the object during measurement and to use video cameras and computer processing of images. For instance, instead of a single "average" value of pH of the whole cell suspension in the cuvette obtained by a classical technique they offer time- and space-resolved maps of various ions inside an individual cell and in its immediate neighbourhood (Fig. 13.2).

The dyes used as fluorescent probes for ionic composition work on the classical principle of litmus red–blue transition. The absorption (and/or fluorescence excitation and/or fluorescence emission) spectrum changes when the dye binds a particular ion. Both forms, free and bound, can be easily spectroscopically distinguished, their concentration ratio determined and, using a standard calibration curve, converted into actual pH, pCa values and the like. That is why these techniques are often called the "ratio method" or "ratio imaging" (if maps of local concentration in the cell and its neighborhood are derived). For some ions no dyes with spectral shifts exist and a selective ion-specific fluorescence quenching must be employed. Dyes with a spectral shift in excitation (dual excitation) are more convenient for microscopy (one camera, two excitation filters) while dyes with an emission shift (dual emission) are often preferred in flow cytometry and cell sorting (one excitation laser wavelength, two detectors).

Though many dyes exhibit some kind of dependence of the absorption and/or fluorescence on the ionic composition of the medium, only a few of them

Fig. 13.1. Fluorescent probe gives "classified" information on the behaviour of surrounding molecules

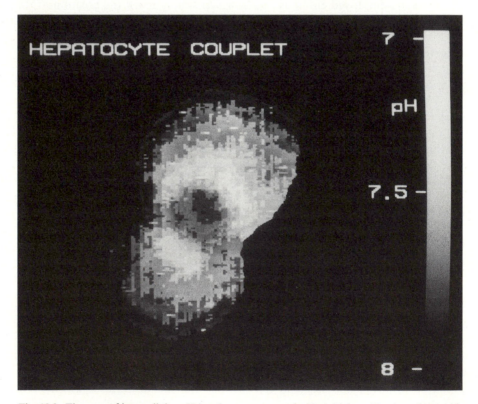

Fig. 13.2. The map of intracellular pH in a hepatocyte couple. Note higher pH values close to the intercellular space (BCECF dye, Hamamatsu SIT camera, Nikon image analysis system LUCIA)

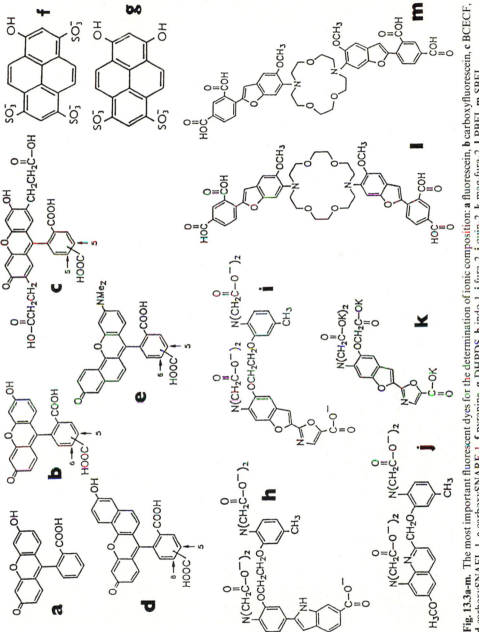

Fig. 13.3a-m. The most important fluorescent dyes for the determination of ionic composition: **a** fluorescein, **b** carboxyfluorescein, **c** BCECF, **d** carboxySNARF-1, **e** carboxySNAFL-1, **f** pyranine, **g** DHPDS, **h** indo-1, **i** fura-2, **j** quin-2, **k** mag-fura-2, **l** PBFI, **m** SBFI

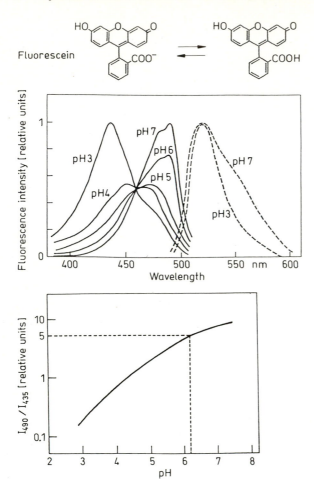

Fig. 13.4. The pH dependence of fluorescence of fluorescein and the resulting calibration curve. Note the two excitation peaks at 435 and 490 nm and the pH-independent emission maximum at 490 nm

can serve as a reliable concentration indicator for a particular ion. Among the best, but still far from perfect, are fluorescein and its derivatives for pH measurement and fura-2 for calcium estimation (Figs. 13.3 and 13.4, Table 13.1).

2 Theory of Color Change

The color change of indicators is explained either as a direct consequence of binding of the ion to the indicator molecule (Ostwald's theory [2]) or as a result of tautomerization following ion binding (Hantzsch's theory [3]), Fig. 13.5. However, the resulting equation of the Ostwald theory is usually amended by introducing additional "empirical constants" and thus it becomes formally quite identical with the resulting Hantzsch equation or with the calcium calibration curve used by Grynkiewitz. On the basis of Hantzsch's theory this means that mere binding or release of an ion represents a change in the indicator molecule too small to cause a color change, and additional steps are expected. For practice this means that all these descriptions are in fact equivalent [1].

Hantzsch theory:

$$pH = \log K_2 + \log \frac{R - K_2/K_1 K_T}{1/K_T - R}.$$

Ostwald theory:

$$pH = pK_{eff} + \log \frac{R - R_{min}}{R_{max} - R} - \log \frac{F_a}{F_b}$$

$$pH = \log(K_{eff} F_b/F_a) + \log \frac{R - R_{min}}{R_{max} - R}.$$

Grynkiewicz:

$$[Ca^{2+}] = K_d \cdot \frac{R - R_{min}}{R_{max} - R} \cdot \frac{F_a}{F_b}.$$

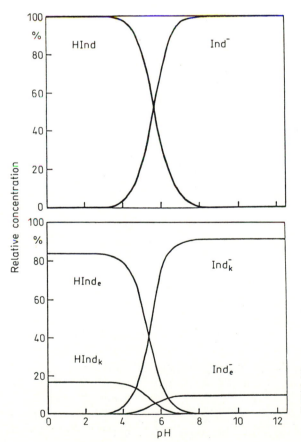

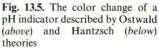

Fig. **13.5.** The color change of a pH indicator described by Ostwald (*above*) and Hantzsch (*below*) theories

Table 13.1. List of dyes

PH probes	Acidic form		Basic form		AM form		pK
	ex	em	ex	em	ex	em	
Derivatives of fluorescein:							
fluorescein	435	520	490	520	nonflourescent		2.4 & 4.6 & 6.7
carboxyfluorescein	435	520	490	520	nonfluorescent		6.4
BCECF	437	520	508	531	nonfluorescent		7.0
fluorescein sulfonic acid	NA	NA	492	520	nonfluorescent		6.4
5,6-carboxydichlorofluorescein	NA	NA	504	529	nonfluorescent		5.1
5,6-carboxydimethylfluorescein	NA	NA	504	535	nonfluorescent		7.0
carboxy SNAFL-1	479 & 508	543	537	623	nonfluorescent		7.8
carboxy SNAFL-2	485 & 514	546	547	630	nonfluorescent		7.65
carboxy SNARF-1	518 & 548	587	579	640	516 & 548	590	7.5
carboxy SNARF-2	517 & 550	583	577	633	518 & 553	588	7.75
carboxy SNARF-6	496 & 524	555	556	635	NA		7.75
carboxy SNARF-X	529 & 569	599	587	630	516 & 548	590	7.9
SNAFL-1	480 & 508	539	537	618	nonfluorescent		7.85
Derivatives of pyranine:							
1-Hydroxypyrene-3,6,8-trisulfonic acid (pyranine)	400	515	460	515	369	397	7.2–7.8[a]
1,3-Dihydroxypyrene-6,8-disulfonic acid (DHPDS)	405	465	458	500	nonfluorescent		7.33 & 8.53
1-Hydroxypyrene-3,6,8-tris-(dimethylsulfonamide)	425	550	495	550	NA		5.7

Calcium Probes

	High Ca^{2+}		Low Ca^{2+}		AM-form		K_d nM
	ex	em	ex	em	ex	em	
fura-2	335	505	362	512	369	478	224
fura-3	341	459	368	462	373	494	140
indo-1	331	410	349	485	361	479	250
quin-2	332	492	352	492	352	492	126

Magnesium probes

	High Mg²⁺		Low Mg²⁺		AM-form		K_d mM
	ex	em	ex	em	ex	em	
mag-fura-2	344	492	376	506	376	506	1.5
mag-indo-1	349	419	354	475	361	467	2.7
mag-quin-1	335	490	349	499	349	NA	6.7
mag-quin-2	337	493	353	487	349	NA	0.8

Sodium probes

	High Na⁺		Low Na⁺		AM-form		K_d mM
	ex	em	ex	em	ex	em	
SBFI	334	525	346	551	396	570	17–19[b]

Potassium probes

	High K⁺		Low K⁺		AM-form		K_d mM
	ex	em	ex	em	ex	em	
PBFI	334	525	346	551	360	512	107[c]

Chloride probes
(only one form of the dye exists, collisional quenching mechanism)

	ex	em	K_q
SPQ	344	450	12 to 118 M⁻¹
SPA	415	488	NA

Data have been compiled from catalogues of Molecular Probes [6] and Lambda Probes and Diagnostics [7].
Wavelengths are in nanometers. AM denotes the membrane permeant ester of the dye used for the loading of cells (precursor)
Abbreviations:
[a] pK depends strongly on ionic strength
[b] K_d depends on the concentration of potassium
[c] K_d depends on the concentration of sodium
NA data not available

3 Calibration Curve

The main problem of the intracellular measurement of ions is how to find the correct calibration curve. The calibration curve is usually determined experimentally either in situ by manipulating the intracellular concentration of ions using ionophores or in vitro in series of buffers of known ion concentration. As can be seen from the above equations [1], any shift in pK results in a proportional shift in the measured value. If the pK is stable, then calibration is simple and reliable. If not, different approaches yield different calibration curves. For instance, for fura-2, where the pK depends strongly on intracellular pH, ionic strength, viscosity and temperature, even intracellular calibrations made in different cell types are different. The usual approach with fura-2 is to adjust the equation coefficients in the calibration equations for an "infinite" calcium concentration (when almost all dye molecules are bound to calcium, i.e. after addition of an ionophore, such as valinomycin or A23187 which triggers influx of extracellular calcium into the cell) and "zero calcium" (after addition of this ionophore together with high concentration of the chelator EGTA, which binds all intracellular and extracellular calcium better than fura-2 does).

To obtain reliable results one should get familiar with all the weak points of the indicator dye. These effects include unspecificity for the measured ion (namely K^+ versus Na^+ or Ca^{2+} versus Mg^{2+}), existence of more than two fluorescent forms (e.g. fluorescein has 10 [1]), photolability, formation of photochemical products which may also fluoresce and bind the ion (e.g. photoproducts of fura-2 bind also calcium, but with different pK's), fluorescence of the precursor (see loading of cells with indicator).

4 How to Bring the Indicator Dye into the Cell

The most popular way of bringing the fluorescent dye into the cell is that introduced by our laboratory in 1982. It uses a permeant precursor, which is hydrolyzed inside the cell to an impermeant form which is trapped inside. The best precursor is nonfluorescent or colorless or both, so that it does not appear in the resulting fluorescence signal (Fig. 13.6).

Other possibilities of loading cells with indicator dye include micro-injection, temporary permeabilization of the cell membrane by mechanical or chemical means, long term incubation under the condition in which the dye bears no electrical charge and thus is membrane-permeant, e.g. low or high pH for pH probes.

In cellular applications, the homogeneity of distribution of the dye inside the cell must be taken into account (e.g. fluorescein penetrates into mitochondria while carboxyfluorescein does not), the toxicity of the dye and the possible direct buffering effect of the dye on the concentration of the ion measured. The usual concentrations of the dye are 10^{-5} to 10^{-7} M, (physiological concentra-

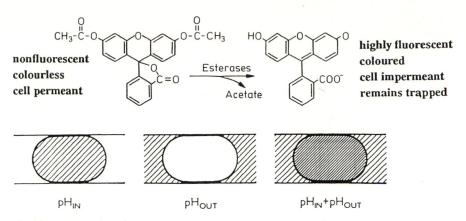

FLUORESCEIN DIACETATE   **FLUORESCEIN**

nonfluorescent
colourless
cell permeant

Esterases

Acetate

highly fluorescent
coloured
cell impermeant
remains trapped

pH_{IN} pH_{OUT} $pH_{IN} + pH_{OUT}$

Fig. 13.6. The loading of cells with an ion indicating dye

tions of calcium or protons are 10^{-8} to 10^{-6} M). However, the decrease of free calcium or proton concentration after the dye addition is readily equilibrated by intracellular buffering effects.

5 How Fluorescence is Measured

The variety of ways how fluorescence is detected reflects the great versatility of fluorescence techniques. The fluorescence signal can be detected from the surface of whole organs in vivo using fiber optics as well as from tissue section, cells grown on coverslips, cell suspensions in cuvettes, selected spots inside the cells, or from individual points of cell images in a microscope. Every approach has its pros and cons.

Measurement on tissues, tissue sections, whole suspensions of cells in a cuvette, or cells on coverslips give one "average" value for all cells (Fig. 13.7). In flow cytometry, the signal from each cell is registered individually and gives one average value for each cell. This value can be also combined with another cell parameter, e.g. size, morphology, an other ion and, in a cell-sorting apparatus, such as Fluorescence Activated Cell Sorting (FACS) used as a parameter, on the basis of which the cells are physically separated (Fig. 13.8).

Spot measurement under a microscope allows one to detect the fluorescence signal from a certain selected area, the size of which can vary from a millimeter to submicrometer, depending on the diaphragm size and the magnification of the microscope. This approach allows correlation with morphology of the cell and is convenient for measurement of fast changes.

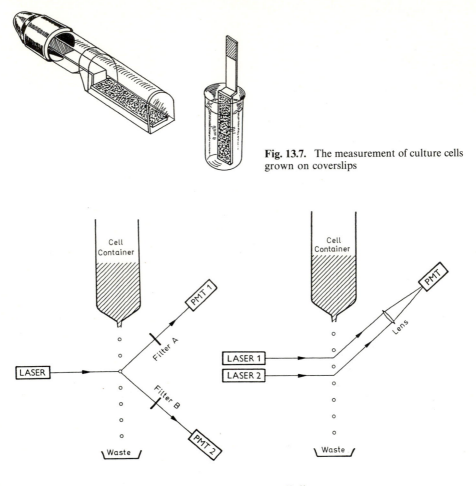

Fig. 13.7. The measurement of culture cells grown on coverslips

Double emission

Double excitation

Fig. 13.8. Flow cytometry and cell sorting

However, the imaging techniques (Fig. 13.9), though more expensive and sometimes slower give incomparably more information than spot measurement does (cf. Fig. 13.2).

Gradients of ions across the membrane can be (either with or without a microscope) measured after addition of two different dyes as follows. Addition of a membrane-permeant precursor which is converted into an ion-sensitive dye and then remains trapped inside the cell, gives information on intracellular ion concentration. The addition of a membrane-impermeant dye gives information on the extracellular ion concentration. The combination of the two dyes, one trapped inside the cell and one outside the cells offers the possibility to measure gradients across the membrane (Fig. 13.6).

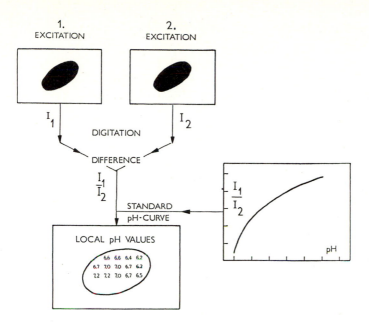

Fig. 13.9. Principle of intracellular imaging

Another possibility of using two dyes is to incubate cells both with a pH sensitive and with a calcium-sensitive dye. Using selective excitation, one can use e.g. SNAFL or SNARF in the region of 580–620 nm for pH measurement and fura-2 around 510 nm for calcium measurement.

6 Limitations

6.1 Temporal Resolution

The relaxation time constants for fluorescent dyes lie in the range of 10^{-3}–10^{-2} s [1, 4]. A spot measurement using a fast revolving filter wheel requires about 20–50 ms per value, processing of images by a fast computer is possible in real time, i.e. 25 pictures per second.

6.2 Spatial Resolution

Spatial resolution is limited by the theoretical limit 150–200 nm for a classical optical microscope and around 100–150 nm for a confocal microscope. The definition of a classical microscope can be improved by special deconvolution mathematical procedures which also remove the fluorescence flare and the degradation of contrast by out-of-focus blurring. Nevertheless, the real confocal

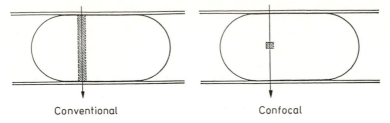

Conventional   Confocal

Fig. 13.10. Limitation in classical and confocal microscopy

microscope (however expensive and somewhat slow or insensitive it may be) yields a much higher contrast due to its "point probing" in which only in-focus information is selected and provides basis for subsequent three-dimensional reconstitution (Fig. 13.10).

6.3 Temporal and Spatial Resolution

There is a trade-off between sensitivity and time resolution. Suppose we have a 1 μM concentration of the dye, which corresponds to only four molecules per 0.2 μm cube. One dye molecule can survive a limited number of excitation-emission cycles before it is chemically decomposed (for fluorescein this is about 10^4 to 10^5 cycles [5]). With respect to the sensitivity of SIT CCD cameras and of the quantum yield of the dye, this limits the number of images that can be captured to several hundreds or thousands, after which the dye is lost. For longer measurement, either larger areas or higher dye concentrations must be used.

7 Conclusion

In spite of all the limitations discussed above, fluorescence dyes represent one of the best and most reliable techniques for the measurement of intracellular ion concentration. Though the absolute values may not be always reliable, very small changes (e.g. 0.01 pH unit) or very fast changes can be easily detected. The imaging possibilities are outstanding. It is a promising technique with a great future.

8 References

1. Kotyk A, Slavik J (1989) Intracellular pH and its measurements, CRC Press, Boca Raton, Florida
2. Ostwald W (1929 and 1937) Grundsätzliches zur Farbenlehre, Akad Wiss (Berlin)
3. Hantzsch A (1918) Berichte 50: 1413
4. Tsien RY (1989) Ann Rev Neurosci 12: 227

5. Waggoner AS (1986) In: Taylor DL et al. (ed) Applications of Fluorescence in the Biomedical Sciences Alan R. Liss, New York, p 3
6. Haugland RP (1989) Handbook of fluorescent probes and research chemicals. Molecular Probes, Oregon, USA
7. Koller E (1989) Innovation in fluorescent probes, Lambda Probes and Diagnostics. Graz, Austria

Part 3. Fluorimetric Analysis

14. Analytical Applications of Very Near-IR Fluorimetry

James N. Miller, Marc B. Brown, Nichola J. Seare and Stephen Summerfield

Department of Chemistry, Loughborough University of Technology, Loughborough, Leicestershire LE11 3TU, UK

1 Introduction

Fluorescence spectrometry has been accepted for many years as a major technique for trace analysis, and it is applied routinely and successfully to such diverse fields as the detection of solutes in flowing systems (e.g. in HPLC, CZE, and FIA); the monitoring of biospecific reactions, as in immunoassays and DNA-probe and enzyme assays; the detection of inorganic ions such as H^+ and Ca^{2+} in cellular and other samples; the detection of components separated by laminar methods such as TLC and zone electrophoresis; and the study of molecular interactions, such as ligand–protein binding. It is noteworthy that in most of these assays it is not the intrinsic fluorescence of the determinant which is measured (polynuclear aromatic hydrocarbons are the major exception to this generalization): in most applications fluorescent labels or probes are used to provide the desired optical and molecular properties. [In this paper the term *label* is used to describe a covalently bound fluorophore group, while *probe* means a fluorophore, usually non-covalently bound to a protein or other surface, whose environment dependent fluorescence properties give information on the polarity of its binding sites]. The major advantages of fluorimetry in such fields are increased selectivity compared with UV-Visible absorption spectrometry, the great variety of sample handling methods available, and most of all the exceptional limits of detection accessible in favourable circumstances.

But the practical limits of detection for many fluorophores are often several orders of magnitude higher (i.e. poorer) in real sample matrices than those obtained in pure solutions of the same fluorophores: this is because of background scattered light and fluorescence signals from endogenous sample components. Such problems are especially severe when biological matrices are studied, as their high content of natural polymers causes intense light scattering signals, and such samples commonly contain numerous fluorophores whose spectra wholly or partly overlap those of the target fluorophore. Even in analytical methods in which the fluorescence determination is preceded by a separation step, background signals due to solvent scattering, trace fluorescent impurities in solvents, cuvettes etc. are often significant.

Recent work in several laboratories has shown that these problems can be successfully tackled by using fluorophores which absorb and emit at higher wavelengths: The useful region for long-wavelength fluorescence emission is probably in the range ca. 650–900 nm: this range includes the extreme red end of the visible spectrum as well as the very near infra-red (VNIR) region, but the latter description is used throughout this paper. The major benefits of VNIR fluorimetry are expected to be much reduced Rayleigh and Raman scattered light signals, because of the inverse fourth-power intensity-wavelength relationship and the very long-wavelength Raman shifts, the reduction in background fluorescence signals because there are so few bright VNIR fluorophores, and the ability to develop simple, robust but sensitive instrumentation based on solid-state light sources and detectors and cheap and effective fibre optics. This paper surveys some recent work in this area, and demonstrates the potential of VNIR fluorimetry for several applications.

2 VNIR Fluorophores

Photochemical theory predicts a sharp diminution in the quantum yields of fluorophores as their emission wavelengths increase [1]. In practice the VNIR region can boast a few fairly intense fluorophores, with almost all other molecules being effectively non-fluorescent. Since, as already noted, most fluorimetric assays are based on label or probe molecules, this is a very favourable situation – to develop a range of assays it is only necessary to synthesize or modify fluorophores to provide suitable reactive groups. Fluorophores with moderate molar absorptivities and quantum yields may give excellent limits of detection because of the low backgrounds. The dye families with VNIR emission are xanthenes, phenoxazines, carbocyanines, and phthalocyanines. All but the last of these groups have been studied in the authors' laboratory.

2.1 Xanthene Dyes

The classical xanthene fluorophore, fluorescein, still finds much application, but its excitation and emission wavelengths are especially unsuitable for work in biological samples such as blood plasma, its emission wavelength corresponding closely to that of protein-bound bilirubin. Longer wavelength xanthene derivatives such as Texas Red (Sulphorhodamine 101 chloride) [2] are also popular, but still longer excitation and emission wavelengths are desirable if the full advantages of VNIR operation are to be obtained. Rhodamine 800 [3] seems to be a good candidate: its major long-wavelength excitation bands are at ca. 635 and 690 nm, but it can be efficiently excited by modern short wavelength diode lasers operating at 650–670 nm, and with an emission wavelength in aqueous solution of ca. 720 nm Rayleigh scattering interference is very small (see below). Like other xanthene dyes rhodamine 800 does not have particularly environment sensitive fluorescence properties, though its fluorescence is enhanced ca. 3-fold in acetonitrile solution compared with aqueous solution. Its interactions with serum albumin and α-1 acid glycoprotein produce only small fluorescence enhancements (ca. 1.3- and 2.5-fold respectively) and blue shifts in emission wavelength (0 and ca. 9 nm respectively). This dye is thus not suitable as a fluorescence probe molecule, but it has potential as an ion-pairing reagent and any derivatives with active functional groups may be extremely useful VNIR fluorescent labels. Similar remarks may apply to derivatives of Vita Blue, synthesised so as to be excitable by He–Ne lasers at 633 nm [4].

2.2 Phenoxazine Dyes

Several phenoxazine dyes have been identified as having potential as both labels and probes, including cresyl violet [5] and Nile Red. The latter dye was originally discovered as an impurity in the well-known long wavelength quantum counter Nile Blue and has been shown to have fluorescence properties which are strongly dependent on the polarity of the solvent [6]. It thus seemed likely to be of value in probing ligand protein interactions, and in the author's laboratory studies have been carried out involving the binding of Nile Red to serum albumin, α-1 acid glyocoprotein, ovomucoid and β-lactoglobulin, and its displacement from these proteins by other ligands (see below). The synthesis of a Nile Red derivative with reactive groups suitable for protein labelling has recently been reported [7]. Another oxazine dye of interest is oxazine 750, available as a laser dye and with an excitation wavelength well matched to 660–670 nm diode lasers [8]. This potential label/probe (along with Nile Blue) has been covalently bound to albumin using a carbodiimide reaction [9] and is currently under further study in our laboratory.

2.3 Carbocyanine and Merocyanine Dyes

The fluorescence properties of several dicarbocyanine and tricarbocyanine dyes have been well characterized: many such derivatives are commercially available, and fluoresce at wavelengths up to 1000 nm. Recent studies have shown their potential as both probes and labels. The tricarbocyanine dye indocyanine green binds non-covalently to albumin and other proteins [10] allowing their detection at pM levels with the aid of a 780 nm diode laser for excitation. Another tricarbocyanine, DTTC [3,3'-diethylthiacarbocyanine perchlorate], has interesting fluorescence properties: in contrast to other probe molecules, its excitation and emission wavelengths *fall* as the polarity of the solvent environment increases (Fig. 14.1). The wavelength changes are small, however, and such cyanines show little promise as polarity probes. In view of the low energy transitions involved there is concern that VNIR fluorescence might show an unusually large temperature dependence. This fear is diminished by our finding that the fluorescence of DTTC falls by only ca. 0.5% per °C in the range 18–45 °C, a rate of change less than that of many UV-Visible fluorophores. Several carbocyanine derivatives have been proposed as covalent labels of proteins etc. via a range of functional groups [11–13] including isothiocyanate and succinimide groups.

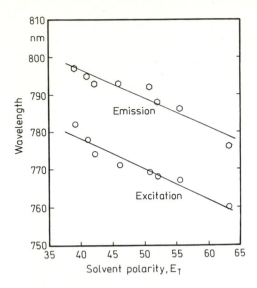

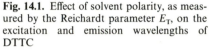

Fig. 14.1. Effect of solvent polarity, as measured by the Reichardt parameter E_T, on the excitation and emission wavelengths of DTTC

In contrast to cyanine dyes, which contain a delocalised positive charge, merocyanines can exist in uncharged or dipolar forms: merocyanine 540 is the best-known example of the latter group of dyes. It has been suggested that such compounds are good probes for solvent polarity [11], but our protein-binding studies showed that only in the case of albumin was the dye–protein interaction accompanied by a significant fluorescence intensity change. As in the case of the carbocyanines, protein binding was accompanied by a small (ca. 9 nm) *red* shift in emission wavelength, i.e. in the opposite direction to the shifts shown by classical probe molecules.

3 Instrumentation

VNIR fluorescence measurements can be performed using a great variety of instrumentation. For many researchers, the simplest approach will be to use a conventional fluorescence spectrometer with a red-sensitive photon multiplier tube. Such instrumentation will be effective provided that the excitation wavelength is not so high that the typical xenon light source and excitation grating monochromator provide very poor incident intensities. A further difficulty is that some conventional quantum counters are ineffective above ca. 700 nm, so ratio recording systems are problematical at such high wavelengths. The use of large conventional fluorescence spectrometers also sacrifices one of the potential benefits of VNIR fluorimetry, the use of simple optical components (see above).

The simplest instruments for VNIR fluorimetry utilise bright light emitting diodes as sources and simple photodiodes as detectors. Such systems are capable of determinations at μM and nM levels (J.N. Miller and D.L. Riley, to be published), and lower limits of detection are probably feasible with avalanche photodiodes as detectors. Multi-colour photodiodes may make multi-component analyses feasible. As in other spectroscopic techniques, a great enhancement of information content can be obtained by using multi-channel detectors. We have made VNIR fluorescence measurements using an intensified multichannel photodetector coupled to an f/4.5 Czerny-Turner monochromator. Fibre optics were used to bring light from a variety of sources to the sample compartment, and to take light from the sample (in a conventional 10 mm cuvette) to the detector. Light sources compared included a 2 mW laser diode emitting (without temperature control) at 677 nm, a 275 W xenon arc lamp and a 50 W tungsten-halogen lamp. Some results obtained using this system are discussed below.

Fluorimetry is frequently used for monitoring flowing systems, and VNIR fluorimetry is no exception. It has been shown in the author's laboratory that 10 ng ml^{-1} levels of rhodamine 800 can be determined in a very simple fluorescence detector fitted with a 35 microlitre flow cell (ca. 1 ng injected), and femtogram limits of detection have been claimed in a system utilising a micro flow cell and a diode laser light source [14]. The applications of VNIR fluorimetry to studies of solid surfaces such as TLC plates, microtiter plates etc. is also very promising in view of the reduced light scattering interference.

VNIR fluorophores can also be readily excited by chemiluminescence energy transfer from (e.g.) peroxyoxalate systems [15]. We have shown that Nile Red excited via the hydrogen peroxide oxidation of TCPO (bis-[trichlorophenyl]oxalate) can be determined at levels as low as 5×10^{-13} M in a simple fluorimeter with its light source extinguished.

4 Results and Applications

The intensified diode array system described above was used principally to study the carbocyanine dye DOTC [3,3'-diethyloxatricarbocyanine iodide]. It was shown that the limit of detection for this compound using diode laser excitation was 1.6 ng ml^{-1}, a value very similar to that obtained with a conventional fluorescence spectrometer, and several times lower than the values obtained using the multi-channel detector with Xe or tungsten-halogen sources. Although the Stokes shift for DOTC is small, < 40 nm in this experiment, the effect of Rayleigh scattered light was negligible because of the narrow bandwidth of the light source. The same system was used to show that another tricarbocyanine dye, DTTC, binds very rapidly to serum albumin with a fluorescence enhancement of ca. 4-fold, and that the fluorescence of the complex then falls by ca. 15% over a period of 600 s (Fig. 14.2). Each point in the figure represents a reading from a complete spectrum of the protein–dye complex: no changes in fluorescence emission wavelength were observed over this period.

A major VNIR study in our laboratory has been the use of Nile Red as a fluorescence probe for ligand binding sites on protein surfaces. This dye has an emission maximum in aqueous solution (pH 3–10) of ca. 665 nm, but it exhibits

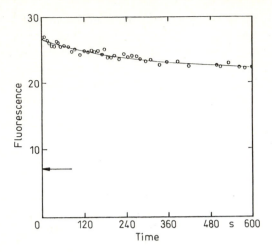

Fig. 14.2. Time dependence of fluorescence at 780 nm of DTTC bound to bovine serum albumin at pH 7. Albumin and DTTC concentrations were 0.3 and 1.8 µM respectively. The *arrow* shows the fluorescence intensity of DTTC in the absence of albumin

Table 14.1. Nile red as a protein probe

Protein	Emission wavelength nm	Fluorescence enhancement
Bovine Serum Albumin	632	35
β-Lactoglobulin	610	100
α_1-Acid Glycoprotein	635	15
Ovomucoid	650	8

significant blue shifts of this maximum, and significant fluorescence enhancements on binding to albumin, β-lactoglobulin (which is believed to bind retinol and other ligands), and α-1 acid glycoprotein and ovomucoid, both of which are of intense current interest because of their ability to bind basic drugs and, where appropriate, resolve their chiral forms (Table 14.1). The value of these phenomena is demonstrated by further studies in which excess D,L-propranolol was added to the Nile Red–protein complexes. No displacement of the dye occurred in the case of β-lactoglobulin, and only a small diminution of Nile Red fluorescence was observed in the case of albumin. In contrast the dye was completely displaced by D,L-propranolol from α-1 acid glycoprotein, and almost completely displaced from ovomucoid: propranolol is known to bind to both of these proteins. In each case the displacement was shown by the quenching of the Nile Red fluorescence, and the reversion of the emission wavelength to ca. 660 nm (Fig. 14.3). Such studies allow the binding constants and numbers of binding sites on the protein surface to be determined. There is also the intriguing possibility that such phenomena might be used to develop simple sensors for the chiral forms of enantiomeric drugs: it was found that the displacements of Nile Red from α-1 acid glucoprotien by D-propranolol and

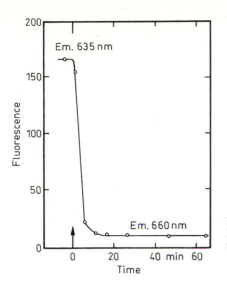

Em. 635 nm

Em. 660 nm

Fig. 14.3. Displacement of Nile Red from a 1:1 Nile Red:alpha-1 acid glycoprotein complex by the addition (at time 0) of D,L-propranolol

L-propranolol were not identical. Further interesting phenomena remain to be studied: for example, it was found that racemic ephedrine caused an 85% reduction in the fluorescence of Nile Red bound to α-1 acid glycoprotein, *but* without a change in the fluorescence wavelength (J.N. Miller, N.J. Seare and M.B. Brown, unpublished).

Other studies of VNIR fluorescence in our laboratory include the development of new covalent labels, and their application to both homogeneous and heterogeneous immunoassays; the use of VNIR fluorophores as ion-pairing reagents in extraction and chromatography; and the development of portable VNIR spectrometers suitable for field use in the determination of two or more materials simultaneously.

5 Conclusions

This brief survey of VNIR fluorimetry and its applications shows its potential as an extremely powerful analytical tool. With the development of new VNIR fluorophores, and further advances in solid state light sources and detector instrumentation, it is not too much to expect that, within a few years, many applications of fluorimetry which currently use UV-Visible fluorophores will use VNIR methods instead, and the availability of very simple, robust, perhaps portable instruments will greatly extend the range of analytical and other applications of fluorescence spectrometry. Crucial to such developments will be the synthesis of new fluorescent labels and probes with good stability and solubility, and carrying the variety of functional groups that is necessary to label and probe the numerous features of bio-organic molecules.

Acknowledgements. Some of the work summarised briefly in this paper has been performed by Analytical Chemistry Research Group members Ravinder Hindocha and David Riley. We also thank Dr. Tony E. Edmonds for valuable discussions, and the following bodies for their financial support: the Science and Engineering Research Council, The Agricultural and Food Research Council, and the Royal Society of Chemistry.

6 References

1. Heitler W (1954) The quantum theory of radiation. Clarendon Press, Oxford
2. Titus JA, Haugland RP, Sharrow SO, Segal DM (1982) J Immunol Methods 50: 193
3. Raue R, Harnisch H, Drexhage KH (1984) Heterocycles 21: 167
4. Lee LG, Berry GM, Chen C-H (1989) Cytometry 10: 151
5. Alvarez-Buylla A, Ling CY, Kirn JR (1990) J Neurosci Methods 33: 129
6. Fowler SD (1985) J Lipid Res 26: 781
7. Petrotchenko EV, Gorelenko AY, Alekseev NN, Kiselev PA, Akhrem AA (1991) Poster Presentation at the 4th International Symposium on Quantitative Luminescence Spectrometry in Biomedical Sciences, Ghent, Belgium, May 1991
8. Aumiller GD (1982) Opt Commun 41: 115
9. Sauda K, Imasaka T, Ishibashi N (1986) Anal Chem 58: 2649
10. Imasaka T, Tsukamoto A, Ishibashi N (1989) Anal Chem 61: 2285
11. Ernst LA, Gupta RK, Mujumdar RB, Waggoner AS (1989) Cytometry 10: 3
12. Mujumdar RB, Ernst LA, Mujumdar SR, Waggoner AS (1989) Cytometry 10: 11
13. Waggoner AS (1990) Ger Offen DE 3 912, 046
14. Kawabata Y, Imasaka T, Ishibashi N (1986) Talanta 33: 281
15. Rauhut MM, Roberts BG, Maulding DR, Bergmark W, Coleman R (1975) J Org Chem 40: 330

15. Fluorescence Detection in Flow Injection Analysis

M. Valcárcel and M. D. Luque de Castro

Department of Analytical Chemistry, Faculty of Sciences, University of Córdoba, 14004, Córdoba, Spain

1 Introduction

According to Taylor [1], a clear distinction should be made between analytical methods and techniques. In doing so, Flow Injection Analysis (FIA) can be regarded as composed of a firmly established series of continuous-flow methodologies developed for introduction of samples and standards into instruments (mainly optical and electrochemical) with or without a chemical derivatization reaction and/or continuous separation. The maturity of FIA is reflected in the large number of research papers (over 3000) and monographs [2–4] published on it, as well as on its steadily increasing use by routine and R&D laboratories. In short, FIA can be considered to be a useful analytical tool for solving a wide variety of analytical problems [5].

The first paper on FIA and fluorimetric detection, which was published in 1978 [6], has been followed by others that describe a number of fluorimetric determinations based on a variety of flow injection manifolds. Most of these configurations rely on the use of conventional manual fluorimetric methods or air-segmented continuous-flow methods adapted for use in flow injection configurations. The admittedly lower sensitivity (10–20%) of FIA compared with steady-state methods is offset by its many advantages, which include enhanced selectivity and rapidity; reduced human participation; ruggedness; reduced sample, standard and reagent consumption; versatility; precision; etc.

A flow injection system can be regarded as an interface between samples (and standards) and a conventional fluorimetric detector intended to facilitate the achievement of the following objectives: (a) automatic introduction: (b) development of one or several chemical reactions; and (c) development of non-chromatographic continuous separation techniques involving chemical reactions or not.

The aim of this paper is to describe the most significant innovations in fluorescence spectroscopic techniques applied to flow injection systems reported in the last few years. The earliest approaches in this context are described in the monographs [2–4], so they are not dealt with here. We shall group novel approaches according to the type of reaction involved, the type of flow injection configuration used, non-chromatographic continuous separation techniques used complementarily, and flow-through (bio)chemical sensors based on the integration of continuous separation, fluorimetric detection and – occasionally – chemical reaction in FI configurations.

2 Reactions Involved

The large variety of chemical and biochemical reactions used in FI systems for the development of fluorimetric methods is the best testimony to the capacity of this technique for implementation of any analytical methodology. A host of reactions of highly variable complexity have been used to improve the analytical features of existing methods.

Thus, such *classical chemical reactions* as those based on the use of *o*-phthalaldehyde (OPA), and lumogallion have been used to develop methods that are applicable to the determination of analytes in real samples with enhanced features such is the case with the determination of total primary amines in seawater and plant nectar [7], with the following advantages over a previous automated version of the method: a higher throughput (over 150 vs 30 samples per hour), a lower detection limit (1×10^{-8} M vs 1×10^{-6} M), higher precision [2% at 1×10^{-6} M vs 6% at 1×10^{-5} M], and, probably, higher accuracy (a calculated calibration correction factor is required by the conventional procedure). Reagent consumption, while higher per unit time in the FIA method, is much lower on a per sample basis. This FIA method has also been adapted for the on-line monitoring of amino acids in culture media [8]. Near-real time monitoring is made possible by the fact that measurements are made under non-equilibrium conditions, which is a major asset of FIA. The selectivity and sensitivity of complex formation reactions such as that between Ga and lumogallion [9] can be readily increased by coupling a continuous separation technique [10] such as liquid–liquid extraction on-line [11]. Redox reactions, usually non-specific (e.g. in the thiamine/thiochrome method for the fluorimetric determination of vitamin B1) are also improved in terms of selectivity and sensitivity by using an on-line liquid–liquid extraction procedure [12].

Classical *redox reactions* have been widely used in FIA to obtain fluorescent products with different purposes. Such is the case with total [13] and individual [14] determination of aflatoxins based on their oxidation in the presence of bromine and with the method proposed by Chung and Ingle for the kinetic determination of ascorbic acid by reaction with chloride and *o*-phenylenediamine. Two kinetic methods (fixed-time and reaction-rate) were developed and compared by using very simple FI manifolds [15]. Multideterminations based on redox reactions have also been developed altering the original FI system slightly. Such is the case with the determination of peroxides (H_2O_2 and CH_3HO_2) with the *p*-hydroxyphenylacetate/peroxide/peroxidase system. Discrimination between the two peroxides relies on the use of a valve-switchable MnO_2 reactor in the sample loading line [16]. Speciation of cerium [as Ce(III)-Ce(IV)] was achieved in a similar way by determining Ce(III) on the basis of its native fluorescence and using an on-line zinc reducing minicolumn for determination of Ce(III) + Ce(IV) [17]. Less general reactions such as catalysis of cyclic condensations [18], formation of ternary complexes [19], metal catalysis [20], or inhibition phenomena in the oxidation of hydrazones [21] have also been used in this context.

Enzyme reactions with the catalyst in immobilized [22] or dissolved form [23] have been widely used in flow systems coupled to fluorimetric detectors.

The use of dehydrogenases involves monitoring the reduced form of the coenzyme (NADH) the concentration of which will decrease or increase depending on whether the reduced form is reactant or product of the enzymatic reaction. The determination of the lactic acid content in extracellular fluid after intracerebral dialysis by using dissolved lactate dehydrogenase [24], and that of ethanol by use of dissolved [25] or immobilized [26] alcohol dehydrogenase (with [26] or without [25] cyclic regeneration of the co-factor) are interesting examples in this respect. Mannitol dehydrogenase was immobilized on poly(vinyl alcohol) beads for the determination of D-fructose based on fluorimetric monitoring of the disappearance of NADH [27] and for that of mannitol in beverages by monitoring of NADH formed [28]. The stability of the enzyme reactor (at least two months) plays a major role, especially when methods are to be used in routine analyses. The total [29] and individual determination of bile acids after chromatographic separation [30] has also been performed by using 3α-hydroxysteroid dehydrogenase immobilized on controlled-pore glass (CPG). Multideterminations based on enzymatic reactions generally involve co-immobilization of several catalysts. Such is the case with the sequential determination of glucose and fructose in foods (fruit juices, yoghourt and dessert powder). The procedure involves co-immobilization of hexokinase (HK) and glucose-6-phosphate dehydrogenase (G6PDH) for the determination of glucose and that of both enzymes plus phosphoglucose isomerase (PGI) for the determination of fructose [31].

When oxidases are the biological catalyst involved, a coupled reaction allows the hydrogen peroxide formed to yield a fluorescent product through a redox reaction. The complexity of the reactions involved varies enormously. Thus, while the determination of lactate in human blood requires the immobilization of lactate oxidase only for hydrogen peroxide to be formed [32], up to four reactors are needed to obtain this product in the determination of adenosine and iosine in blood plasma: adenosinedeaminase (ADA) for conversion of adenisine into iosine, purine nucleoside phosphorylase (PNP) for that of iosine into hypoxanthine, xanthine oxidase (XOD) for that hypoxanthine into xanthine and uric acid, and urate oxidase (UOD) for that of uric acid into allatoin [33]. In both cases [32, 33] the H_2O_2 is monitored fluorimetrically after reacting with 3-(*p*-hydroxyphenyl) propionic acid catalysed by horseradish peroxidase (HRP) which was included in the last enzyme reactor of the FI system. The enzymatic activity of oxidases in serum has also been monitored fluorimetrically using auxiliary immobilized enzymes as well. Guanase requires the use of three reactors containing a single auxiliary enzyme each (XOD, UOD, and HRP [34]). The indicator reaction, catalysed by HRP, is developed with the aid of 3-(*p*-hydroxyphenyl) propionic acid.

Although CPG is the most commonly used support for enzyme immobilization, poly(vinyl alcohol) beads [27, 28] have been used as supports for these catalysts. Less usual, but also interesting is the use of tubular microporous membranes [36] and Nafion [37] for entrapping enzymes.

Immunochemical systems have been developed in FI manifolds coupled to fluorimetric detectors for a few years now [38], even with the aid of lasers for inducing fluorescence [39]. The present boom of immunoassays is also reflected in the large number of interesting applied methods reported in the FIA literature. By way of example, the method developed by Locascio-Brown et al. for quantitation of theophylline based on a liposome FI immunoassays system is commented on below. The system components include liposomes that contain fluorophores in their aqueous sites and an immobilized antibody reactor column. Automated analyses are performed at room temperature with picomole sensitivity and a day-to-day coefficient of variation of less than 5% for aqueous solutions. The immunoreactor can be regenerated hundreds of times over 3 months of continuous use, with no measurable loss of antibody activity. With minor changes in assay forms, this procedure can also be used for the quantitation of anti-theophylline [40]. One other FI immunoanalysis method was reported for the fluorimetric on-line monitoring of monoclonal antibodies in the course of a hybridoma cell fermentation. Immunoreactors were developed containing either membranes or magnetic particles as the solid phase for the bound antibodies. The detection range of the different procedures was between 1 and 100 μg/ml mouse immunoglobulin G, which is the range of interest in hybridoma cell fermentation [41].

Microemulsions provide an ideal interface for reactions between a water-soluble reagent and a non-aqueous sample, which can solubilize and concentrate certain ions and molecules and also act by modifying chemical equilibria, reaction rates and reaction pathways [42]. Some of these assets have been exploited in FIA-fluorimetric detection to improve determination methods. The ideal interface for reactions between a water-soluble reagent and a non-aqueous sample provided by microemulsions has been used for the fluorimetric determination of C_6–C_{10} primary amines by derivatization with OPA and 2-mercaptoethanol [43], and for the solubilization of ternary complexes of terbium(III) [44], europium(III) [45], and samarium(III) [46]. The formation of the aluminium/8-hydroxyquinoline-5-sulphonic acid (HQS) complex is very fast in a micellar medium, so it allows the continuous determination of aluminium in flow systems. The method is very sensitive since the complex forms a strongly fluorescent compound in cationic micelles of cetyltrimethyulammonium bromide (CTAB). Interference is greatly decreased compared with the batch method [47].

A very unusual reagent in FI systems is light. The properties of ultraviolet radiation have been exploited to obtain fluorescent products from three phenothiazine compounds: chlorpromazine, promethazine and perphenazine. The photochemical reactions involved were studied and compared with chemical oxidation processes. After the working conditions were optimized, these reactions were applied to the determination of the analytes by stopped-flow injection methods (irradiation in the flow-cell) and normal flow-injection methods (irradiation of the reaction coil). The determination limits thereby achieved were 20 ng/ml (chlorpromazine and perphenazine) and 50 ng/ml (promethazine), with RSD less than 2% in all instances. Analyses for these compounds in pharmaceutical preparations were also carried out [48]. A simultaneous method for the determination of chlorpromazine and promethazine based on photochemical reaction has also been reported. Discrimination between the two analytes is based on the pH difference between the media where the photochemical conversion of each phenothiazine into a fluorescent product takes place [49].

3 Types of Manifold

The chemical system involved in the derivatizing reaction and the features of the sample matrix determine the design of the most suitable manifold for the application of each method. From the simplest single-channel manifolds used for circulation of each chemical system or for measuring the intrinsic fluorescence of the analyte to the most complex designs based on open-closed circuits, nested loop systems or coupled on-line separation techniques, the general schemes of these manifolds are depicted in Figs. 15.1–15.5 in order of increasing complexity.

Single-channel manifolds such as that shown in Fig. 15.1a have been used for the determination of total primary amines with [43] or without [7] the aid of a micellar medium, and by reaction with OPA, the cyclic condensation of malonic acid with acetic anhydride in non-aqueous media, which is the basis for the determination of tertiary amines that act as catalysts [18], and a method for nucleic acid determination by using a commercial fluorochrome [50] has also been developed in single-line FI systems. Even the method proposed by Chung and Ingle for detection and correction of multiplicative interferences in fluorimetry was developed in a single-line FI system [51]. The method is based on

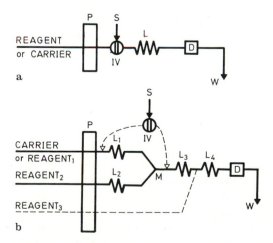

Fig. 15.1. a Simple single-channel and b multichannel, FI manifolds for implementation of fluorimetric methods of different complexity. P denotes peristaltic pump, S sample, IV injection valve, L reactor, D detector and W waste

obtaining a reference peak profile due to physical dilution by injecting a quinine sulphate (QS) solution and comparing it with peak profiles of sample solutions to evaluate the effect of concentration-dependent interference. The distortion of the peak profile due to multiple interferences is detected from a variation in the ratio of the reference-to-sample signals thorughout the sample zone. An empirical function involving a normalization for different analyte concentrations was developed and used to generate correction plots based on the profile distortion. The method was applied to the determination of QS with potassium dichromate (absorber) and KI (quencher) as interferents. Errors due to these interferents as small as 0.5% were detected, and errors larger than 30% were reduced to less than 2%.

Two-channel and, in general, multichannel manifolds (Fig. 15.1b), are used mainly for the following purposes: (a) injecting the sample into an inert carrier which is then merged with the reagent, thus ensuring a better sample-reagent mixing which favours reaction development and increasing sensitivity [8, 44, 46]; (b) mixing buffer and reagent when the pH of the former is appropriate for the development of the reaction but not for reagent stability [52]. In situ mixing minimizes reagent decomposition; (c) allowing implementation of an overall reaction that involves two or more steps. Optimal development is achieved in this case when the steps are developed one by one, the reacting plug meeting each reagent as required by the chemical step [9, 20, 21, 39].

When several steps are involved in the derivatizing reaction or when the reagents to be used are expensive (especially catalysts such as enzymes) there are several ways of simplifying the FI manifold and saving reagents, thereby reducing analytical costs.

Enzymes can be immobilized on a suitable support to construct and enzymatic reactor with substantially reduced catalyst consumption (Fig. 2A). One [27–29], two [32], three [34], four [35], and up to five [33] enzyme reactors can arrange serially in an FI manifold if required by the number of steps involved in the enzymatic reaction. Occasionally, enzymes are co-immobilized and subsequently packed into a single reactor. Such is the case with the determination of glucose and fructose, for which HK and G6PDH, or HK, G6PDH and PGI, respectively, are co-immobilized on CPG [31]. The sequential determination of both analytes can be achieved by using the configuration outlined in Fig. 15.2b, in which a selecting valve allows passage of the injected plug through $IMER_1$ or $IMER_2$, thus allowing the determination of glucose and fructose, respectively.

Redox reagents have also been used in packed form, in reactors which avoid the typical dilution arising from their use in solution. Cerium speciation has been carried out by using a manifold with splitting and merging points quite similar to that depicted in Fig. 15.2b. There is no selecting valve in it, so the sample is split into two channels, one of which includes a zinc reducing minicolumn. A two-peak recording per injected sample is obtained, one peak is due to the sub-plug that is allowed to pass through the column and provides the

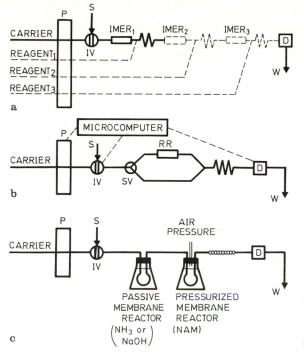

Fig. 15.2a. FI manifold for development of methods involving several consecutive enzymatic reactions; **b** manifold with splitting and merging points for simultaneous or sequential determinations; **c** use of pressurized or passive membranes for introducing reagents into FI systems. *IMER* denotes immobilized enzyme reactor, *SV* selecting valve and *RR* redox reactor (other symbols as in Fig. 15.1)

total content of cerium in the sample, while the other is passed through the open reactor and yields a peak to which only Ce(IV) contributes [17].

Another way of introducing reagents into FI systems is through pressurized or passive membranes. Figure 15.2c shows a single-channel manifold used for the determination of thiols, sulphides and sulphites by reaction with N-acridinyl-maleimide (NAM) in a basic medium (NH_3 or NAOH). A redox reactor (MnO_2) and a tubular microporous membrane entrapped enzyme reactor, together with a passive cation exchange membrane for NH_4OH, was also used by Dasgupta et al. [36], who also employed a manifold with an MnO_2 reactor, a passive ion-exchange membrane for NH_4OH introduction and a pressurized membrane for enzyme introduction in the determination of aqueous peroxides at the nano-molar level [16].

Two or more injection valves can be coupled in several ways for different purposes.

The merging-zones mode (symmetric and asymmetric), which allows the simultaneous determination of several species, with reduced reagent consumption, is based on the coupling in parallel of two injection valves followed by a merging point (Fig. 15.3a). By controlling the propulsion and injection systems through a microcomputer or timer and acquiring and processing fluorescence

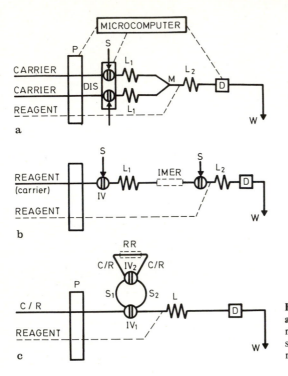

Fig. 15.3a–c. Different valve couplings: **a** in parallel, **b** in series, **c** internal arrangement. DIS denotes dual injection system, *M* merging point, and *C/R* carrier or reagent

intensity-time data with the aid of a microcomputer, one can develop reaction-rate methods with no interference from the sample matrix (e.g. the enzymatic method for the determination of ethanol in whole blood). In addition, the merging point (Fig. 15.3a). By controlling the propulsion and injection systems dissolved enzyme into the manifold with minimum catalyst consumption [25]. The two injection valves of the configuration can be used for sample insertion, as in the simultaneous determination or urea and ammonia in an asymmetric configuration in which the merging point is preceded by an **IMER** containing urease immobilized on CPG. The two simultaneously injected sample plugs reach the fluorimetric detector at different times as each travels through reactors of different geommetrical features before merging (one passes through the enzymatic reactor while the other passes through an open reactor). The derivatizing reagent (OPA) is merged with the main line after **M** [54].

Serial couplings (Fig. 15.3b) have been used together with fluorescence detection for two main purposes: To develop reaction-rate methods without stopping the flow [15], and for simultaneuous determinations [54]. In both cases, the samples were simultaneously injected via the two valves. The kinetic-fluorimetric determination of ascorbic acid based on a double-point method uses a carrier system containing mercuric chloride and *o*-phenylenediamine into which the two sample plugs are injected. The delay time after which both reach the detector gives rise to a two-peak recording. The peak height and area are used as analytical signals. The simultaneous determination of ammonia and urea was performed by placing an urease reactor between the two valves, followed by a point of merging with OPA solution to form the fluorescent product formed with the ammonia from the sample (this plug does not pass through the IMER) and that yielded by the enzymatic reaction (second plug reaching the detector) [54].

Nested loop systems involving internally coupled valves (Fig. 15.3c) have been used in simultaneous determinations as a means of creating pH gradients or splitting sample plugs into two, one of which is not pass through a redox reactor [55]. In the former case, the determination of ammonia and hydrazine was achieved by creating zones of different pH as the analytes reacted with OPA at different pH values. Through the single-line manifold was circulated the reagent in an acid medium, while the loop of IV_2 was filled with the reagent in a basic medium. The sample was used to fill the two sub-plugs into which IV_2 divided the loop of IV_1. By simultaneously switching both valves, sample reagent four interfaces were formed which allowed a four-peak recording to be obtained as the zone reaches the detector. In the simultaneous determination of organic and inorganic peroxides, the valve system allows an MnO_2 reactor to be placed in the loop of IV_2, while the sample is used to fill the two sub-plugs of IV_1. Of these, only the contents of S_1 is passed through the redox reactor, where H_2O_2 is destroyed. Thus, of the two-peaks in the recording obtained on each switching of the valve system to its emptying position, the first corresponds to the sum of organic and inorganic peroxides and the second to organic peroxides only [55].

Cyclic configurations provide special advantages which endow analytical methods with interesting features.

One cyclic system such as that depicted in Fig. 4A was used by Massom and Townshend for the enzymatic determination of ethanol with cyclic regeneration of the coenzyme. In this system, a solution containing buffered NAD^+ passes through a minicolumn of yeast alcohol dehydrogenase immobilized on CPG. The NADH formed is monitored fluorimetrically in the flow system before reconversion to NAD^+ in a minicolumn of glutamate dehydrogenase immobilized on CPG in the presence of glutamate and ammonium ions, also present in the flowing solution. The solution is returned to the reservoir. The regeneration of NAD^+ allows the same coenzyme solution to be used for 50 ethanol determinations per day for 4 days.

An open-closed configuration was used for the determination of sulphate (Fig. 4B) based on the formation of a ternary complex with biacetyl monoxime nicotinylhydrazone (BMNH) and zirconium, the performance of which was compared with that of a normal FI manifold. The repeatability and linear range of the calibration graphs were better for the open-closed that for the nFI system (RSD 2.33% vs 3.52%, and linearity 1.5–150.0 µg/ml vs two linear portions between 2.0–30.0 and 30.0–150.0 µg/ml), but the sampling rate of the nFI configuration (30 samples/h) was at least twice that of the open-closed system [19].

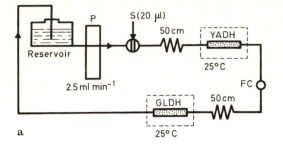

a

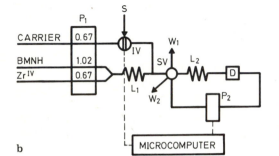

b

Fig. 15.4a, b. Cyclic configurations. **a** closed, and **b** open-closed mode. *YADH* denotes yeast alcohol dehydrogenase, *GLDH* glutamate dehydrogenase, *FC* flow-cell and *BMNH* biacetyl monoxime nitonylhydrazone

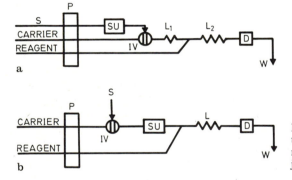

a

b

Fig. 15.5a, b. Location of the separation unit (US) in the flow injection manifold. **a** before and **b** after the injection valve

The on-line coupling of separation techniques with a FI manifold endows methods with enhanced analytical features such as selectivity and sensitivity. Several types of continuous separation techniques have so far been coupled to FI systems. The location of the separation unit in the FI system is dictated by the purpose: in the sample stream prior to the loop of the injection valve (Fig. 15.5a) to remove interferents from the sample matrix, and between the injection and detection system (Fig. 15.5b) for a variety of purposes. Generally, continuous systems are used in post-column derivatization under the name of FI systems. In these, a suitable reagent is continuously merged with the eluate from the column; there are no injection valves other than the high-pressure injection valve of the chromatograph, so, it is not an FI, but a continuous-flow system. Some interesting representative examples of methods including separation techniques and fluorimetric detection are commented on in the next section.

4 Continuous Separation Techniques Coupled to FI Systems

The separation technique most frequently coupled to FI manifolds is liquid–liquid extraction.

A method method for determination of potassium based on the formation of a cationic macrocyclic complex which was extracted into 1,2-dichloroethane as an ion-pair with anilino-naphthalene sulphonate without phase separation was reported in the earlier FI literature [5]. The fluorescent product formed was monitored by passing both immiscible phases through the detector. The method yielded irreproducible results, so there have been some recent successful attempts at improving it by using no phase separation and monitoring one of the phases only [56, 57]. Several other methods based on liquid–liquid extraction with phase separation have been proposed, all of which use a separation unit between the injection and detection system (Fig. 15.5b). Thus, the thiochromo method for the determination of vitamin B_1 in pharmaceutical preparations was adapted to an FI system. The consumption of organic phase was 2–3 ml/sample and the sampling rate was 70 h^{-1}, which an RSD of about 1% [12]. Also, gallium was determined fluorimetrically with lumogallion by an extraction-FI procedure using laser excitation [11], and a multistage FI-solvent extraction system was successfully assayed [58].

Rapid, specific measurement of the lactic acid content in the extracellular fluid of a conscious, freely-moving rat can be accomplished by direct combination of intracerebral dialysis and FIA with enzymatic/fluorescence detection. Basal levels of lactic acid, the end product of anaerobic glucolysis, can be measured by using lactate dehydrogenase to convert lactate into pyruvate, the resulting NADH being the monitored product. The rapid response time of this system, 40 s, allows real-time measurement of transient changes in anaerobic metabolism in the rat striatum [24].

A recent paper by Prinzing et al. reported the use of a less common separation technique (pervaporation) coupled on-line with an FI manifold for bioconversion monitoring [59]. Volatile components of the culture medium such as ethanol or diacetyl are able to diffuse through a hydrophobic membrane driven by a temperature difference. The advantage of the custom pervaporation module designed by the authors was provided by an additional air gap between the process fluid and the membrane, which prevented contact between both. The volatile compounds diffused through the membrane into an acceptor stream and were subsequently driven to the detection system after merging with appropriate reagents.

There are two general approaches to HPLC-FIA couplings that warrant description:
(a) The use of HPLC-FIA systems for on-line monitoring of bioconversion processes [60] as in the on-line enzyme assay proposed by Stammand and

Kula, which allows NAD-dependent oxidoreductases to be monitored during the purification of microbial crude extracts or partially purified enzymes by fast protein liquid chromatography in near-real time. The eluate from the column flows first through a conventional UV detector and then through the injection valve of the analytical system before entering the fraction collector. Assuming plug flow is prevalent in the small-bore tubing used, the delay time between the injection valve and fraction collector is estimated to be 16 s. The injection valve allows the reproducible injection of 2 µl samples into the reagent stream, which consists of a suitable buffer, the substrate and the coenzyme. The signal from the fluorimetric detector is recorded together with the signal from the UV monitor, fraction collector and gradient mixer.

(b) An FI system coupled with a liquid chromatograph can be very useful for routine analyses as it can act both as a screening system for compounds with similar chemical features and as a post-column derivatizing system for the individual determination of the compounds before they are separated. This is the case with the overall and individual determination of aflatoxins in foodstuffs. Injection of the sample through the low-pressure injection valve of the FI system allows a fast screening of the total content of aflatoxins in the sample after a redox reaction with bromine to yield fluorescent products. Then, only those samples with aflatoxins levels close to or above the permissible level are passed through the chromatographic column to determine the amount of each aflatoxin which features a different degree of toxitity [14]. A similar system for bile acids in serum allows the HPLC system to be used to determine the content in each of these analytes in such individuals to facilitate diagnosis [29, 30].

Recently, an FI analytical affinity chromatography system was developed for the on-line monitoring of mouse immunoglobulin G (IgG). Protein A or anti-mouse IgG antibodies immobilized on oxirane beads were packed in a miniature column. The IgG-containing samples and standards were passed through the column and detected fluorimetrically after elution with citrate buffer. Mouse IgG 2a was monitored over a 7-day cultivation of hybridoma cells in a perfusion reactor by FIAAC [61]. A chemical barrier consisting of two three-way magnetic valves and a flask containing 1 M sodium hydroxide was used to prevent contamination by the FIAAC.

5 (Bio)Chemical Fluorimetric Sensors in FI Systems

Two general types of (bio)chemical fluorimetric sensors have so far been used in flow injection manifolds, namely probe and flow-through sensors.

The main shortcoming of probe sensors is the slow response arising from the mass transfer through the membrane. This drastically constrains their use in flow systems. A number of chemically sensitive media have been investigated by

Li and Narayanaswamy for suitability for the development of an oxygen-sensitive optical fibre sensor. The reagent phase was the heart of the transducer, where oxygen was measured by collisional quenching of the immobilized fluorescence indicator by oxygen. Fluorescence quenching is a rapid and reversible process and can be appropriately incorporated into an optical-fibre oxygen transducer. The analytical system employed for measuring the fluorescence response to oxygen and the most suitable reagent matrix were exhaustively studied, but no application was reported [62]. This principle was also used to develop a glucose biosensor based on an oxygen optrode with immobilized glucose oxidase (GOx). Oxygen uptake was determined via dynamic quenching of the fluorescence of an indicator by molecular oxygen. GOx was adsorbed on a sheet of carbon black and cross-linked with glutaraldehyde. Carbon black was used as an optical insulator to protect the optrode from the interference of ambient light and the sample fluorescence. The system provided a linear response for 0.1–500 mM glucose, with an RSD of 2% at 100 mM. The system was applied to the determination of glucose in wine and fruit juice. Up to 60 samples per hour could be analysed and the enzyme optrode remained stable for more than 400 h in continuous use [63]. A very different principle is the basis for the fully reversible fibre optic glucose biosensor proposed by Trettnak and Wolfbeis [64]: the intrinsic green fluorescence of GOx is used to provide the analytical information since the fluorescence of GOx changes during interaction with glucose. The fluorescence is excited at 450 nm and measured at 500 nm, a wavelength range compatible with glass and plastic fibres. The signal response is fully reversible because oxygen is a secondary substrate. A major feature of this sensor lies in the fact that the recognition element is identical with the transducing element. Enzyme solutions are entrapped at the fibre end within a semipermeable membrane, but the change in fluorescence occurs over a small glucose concentration range (typically 1.5–2.0 mM), with a response time of 2–30 min and regeneration times of 1–10 min. Kinetic measurements are suggested as a means of widening the analytical range (2.5–10.0 mM) and shortening response times.

Flow-through (bio)sensors [65] are an inexpensive means of integrating reaction and detection with minimum instrument costs since any ordinary fluorimeter can be used for this purpose and the cell in which the support is placed is usually a commercially available flow-cell.

One peculiar flow-through sensor is that proposed for the photochemical spectrofluorimetric determination of phenothiazines based on irradiation of the flow-cell, which, in which both reaction and detection are integrated by using such an uncommon "reagent" as light [48].

A general approach to the development of luminescent sensors that are selective for a number of cations was recently reported [66]. An ionophore (I) selective to the cation (C) of interest is non-covalently immobilized on CPG particles and a small glass capillary is then filled with the I-CPG phase and placed in the cell compartment of a fluorimeter. As an aqueous mobile phase containing C and 8-anilino-1-naphthalenesulfonic acid (ANS) is pumped through the capillary, the C binds reversibly to the immobilized I and the resulting complex then forms an ion-pair on the CPG surface with the negatively charged ANS. The ANS fluorescence signal, which is highly quenched in the aqueous solution, increases dramatically for the ANS bound to the hydrophobic PCG surface and can be used to determine the amount of C originally present in the mobile phase. Also, the immobilization of the reagent (morin) on a resin exchanger placed in the flow-cell has been used for the determination of a cation (beryllium). The determinative method developed with this sensor features a linear range between 1 and 40 ng/ml, and RSD of 17.1% and a sampling frequency of 30 h^{-1}. Its selectivity allows the determination of the analyte in simulated alloy samples with excellent results [67].

Immobilized reaction products on an anionic resin placed on the flow-cell have been used for the determination of such anions as cyanide and fluoride. The method for the determination of cyanide is based on its reaction with pyridoxal-5-phosphate. The merging-zones FI manifold used allows the simultaneous injection of sample and reagent, the reaction to develop along the transport system and its product to be retained in passing through the flow-cell. The calibration curve obtained is linear over the range 50 ng/ml–3.0 µg/ml, and the RSD and sampling frequency achieved are 1.39% and 10 h^{-1}, respectively [68]. Even more interesting is the flow-through sensor proposed for the determination of fluoride traces in real samples based on the formation of the Zr(IV)-Calcein Blue-F$^-$ ternary complex [69]. A comparison of this sensor with a probe-type sensor [70] based on the same chemical system revealed the flow-through sensor to be clearly superior: determination range 1–20 ng/ml vs 0.5–8.0 µg/ml; detection limit 1 ng/ml vs 0.5 µg/ml; preparation time less than 15 min vs overnight; measurement step 1 min vs 30 min; regeneration step 1 min vs 60 min; reproducibility (RSD) 1% vs 15%, lifetime more than 100 vs 15 measurements.

A multisensor for the determination of pyridoxal, pyridoxal-5-phosphate and pyridoxic acid based on derivative synchronous spectrofluorimetric measurements has also been reported. The determination is based on the formation of fluorescent complexes between the analytes and beryllium in ammonia buffers of different pH values that make it possible to discriminate between pyridoxal and pyridoxal-5-phosphate as the fluorescent features of the two complexes in a basic medium (9.9) are rather similar, but quite different at pH 7.9. Calibration curves between 0.05–15.0 µM were obtained, i.e. higher by two orders of magnitude than those of the conventional flow-injection method and with an RSD smaller than 4.5%. The method was applied to the determination of the analytes in serum samples with good results [71].

6 Final Remarks

Developments in flow injection configurations coupled with conventional fluorimetric detectors furnished with flow-cells have proved this association to be a very promising approach resulting in clearly improved basic analytical features such as sensitivity, selectivity, precision, rapidity, automation and miniaturization capability, sample and reagent consumption, and analytical costs.

The FIA-fluorimetric detection coupling can be used in routine control laboratories to replace conventional approaches and in research laboratories to develop new methods for solving both already existing and emerging analytical problems.

It is also interesting to note that a large number of determinations can be implemented by flow injection fluorimetry in clinical chemistry, food and pharmaceutical analyses, environmental control, etc. Two of the topics dealt with in this paper (use of non-chromatographic continuous separation techniques and flow-through fluorimetric sensors) are probably the most promising trends in this context.

7 References

1. Taylor J (1983) Anal Chem 55: 600A
2. Valcárcel M, Luque de Castro MD (1987) Flow injection analysis: principles and applications, Ellis Harwood, Chichester.
3. Ruzicka J, Hansen EH (1988) Flow injection analysis, 2nd edn. Wiley, New York
4. Karlberg B, Pacey G (1989) Flow injection analysis. A practical guide. Elsevier, Amsterdam
5. Kina K, Shiraishi K, Ishibashi N (1978) Talanta 25: 295
6. Valcárcel M, Luque de Castro MD (1990) Fresenius J Anal Chem 337: 662
7. Petty RL, Michel WC, Snow JP (1982) Anal Chim Acta 142: 229
8. Nalbach U, Schiemenz H, Stamm WW, Kula MR (1988) Anal Chim Acta 213: 55
9. Ishibashi N, Kina K, Goto Y (1980) Anal Chim Acta 114: 325
10. Valcárcel M, Luque de Castro MD (1991) Non-chromatographic continuous separation techniques. Royal Society of Chemistry, Cambridge
11. Imasaka T, Harada T, Ishibashi N (1981) Anal Chim Acta 129: 195
12. Karlberg B, Thelander S (1980) Anal Chim Acta 114: 129
13. Lázaro F, Luque de Castro MD, Valcárcel M (1988) Fresenius Z Anal Chem 332: 809
14. Lázaro F, Luque de Castro MD, Valcárcel M (1988) J Chromatogr 448: 173
15. Chung HK, Ingle Jr JD (1991) Talanta 38: 355
16. Hwang H, Dasgupta PK (1986) Anal Chem 58: 1521
17. Al-Dowdani KH, Townshend A (1986) Anal Chim Acta 179: 469
18. Whiteside IRC, Worsfold PJ (1987) Anal Chim Acta 192: 77
19. Fernández-Band B, Linares P, Luque de Castro MD, Valcárcel M (1991) Analyst 116: 305
20. Lázaro F, Luque de Castro MD, Valcárcel M (1984) Analyst 109: 333
21. Lázaro F, Luque de Castro MD, Valcárcel M (1985) Fresenius Z Anal Chem 321: 467
22. Ruz J, Lázaro F, Luque de Castro MD, Valcárcel M (1988) J Autom Chem 10: 15
23. Linares P, Luque de Castro MD, Valcárcel M (1985) Rev Anal Chem 8: 229
24. Kuhr W, Korf J (1988) Anal Chim Acta 205: 53
25. Fernández-Gómez A, Ruz-Polonio J, Luque de Castro MD, Valcárcel M (1985) Clin Chim Acta 148: 131
26. Masoom M, Townshend A (1986) Anal Chim Acta 195: 49
27. Kiba N, Inouee Y, Furusawa M (1991) Anal Chim Acta 243: 183
28. Kiba N, Inouee Y, Furusawa M (1991) Anal Chim Acta 244: 105
29. Membiela A, Lázaro F, Luque de Castro MD, Valcárcel M (1990) Fresenius Z Anal Chem 338: 749
30. Membiela A, Lázaro F, Luque de Castro MD, Valcárcel M (1991) Anal Chim Acta 249: 461
31. Linares P, Luque de Castro MD, Valcárcel M (1987) Anal Chim Acta 202: 199
32. Zaitsu K, Nakayama M, Ohkura Y (1987) Anal Chim Acta 201: 351
33. Hayashi Y, Zaitsu K, Ohkura Y (1986) Anal Chim Acta 186: 131

34. Hayashi Y, Zaitsu K, Ohkura Y (1987) Anal Chim Acta 197: 51
35. Zaitsu Y, Yamagishi K, Ohkura Y (1988) Chem Pharm Bull 36: 4488
36. Hwang H, Dasgupta PK (1987) Anal Chem 59: 1360
37. Wang J, (1990) Anal Chim Acta 234: 41
38. Kelly TA (1981) Diss Abst Intern 42: 2356B
39. Kelly TA, Christian GD (1982) Talanta 29: 1109
40. Locascio-Brown L, Plant AL, Horváth V, Durst RA (1990) Anal Chem 62: 2587
41. Stöcklein W, Schmid RD (1990) Anal Chim Acta 234: 83
42. Pwlizzetti E, Pramauro E (1985) Anal Chim Acta 169: 1
43. Mwmon MH, Worsfold PJ (1986) Anal Chim Acta 183: 179
44. Ahira M, Arai M, Tomitsugu T (1986) Anal Lett 19: 1907
45. Ahira M, Arai M, Tomitsugu T (1986) Analyst 111: 641
46. Ahira M, Nakashimada T (1989) J Flow Injection Anal 6: 128
47. García-alonso JI, López-García A, Sanz-Medel A, Blanco-González E, Ebdon L, Jones P (1989) Anal Chim Acta 225: 339
48. Chen D, Ríos A, Luque de Castro MD, Valcárcel M (1991) Analyst 116: 171
49. Chen D, Ríos A, Luque de Castro MD, Valcárcel M (1991) Talanta 38: 1227
50. Murray M, Paaren HE (1986) Anal Biochem 154: 638
51. Chung HK, Ingle Jr JD (1990) Anal Chem 62: 2541
52. Hernández-Torres MA, Khaledi MG, Dorsey JG (1987) Anal Chim Acta 201: 67
53. Dasgupta PK, Yang HC (1986) Anal Chem 58: 2839
54. Izquierdo A, Linares P, Luque de Castro MD, Valcárcel M (1990) Fresenius Z Anal Chem 336: 490
55. Dasgupta PK, Hwang H (1985) Anal Chem 57: 1009
56. Cañete F, Ríos A, Luque de Castro MD, Valcárcel M (1988) Anal Chem 60: 2354
57. García-Mesa JA, Linares P, Luque de Castro MD, Valcárcel M (1990) Anal Chim Acta 235: 441
58. Rossi TM, Shelly DC, Warner IM (1982) Anal Chem 54: 2056
59. Prinzing U, Ogbomo I, Lehn C, Schmidt HL (1990) Sensors & Actuators B1: 542
60. Stamm WW, Kula MR (1990) J Biotechnol 14: 99
61. Stöcklein W, Jäger V, Schmid RD (1991) Anal Chim Acta 245: 1
62. Li PYF, Narayanaswamy R (1989) Analyst 114: 663
63. Dremel BAA, Schaffar PH, Schmid RD (1989) Anal Chim Acta 225: 293
64. Trettnak W, Wolfbeis OS (1989) Anal Chim Acta 221: 195
65. Valcárcel M, Luque de Castro MD (1990) Analyst 115: 699
66. Werner TC, Cummings JG, Seitz WR (1989) Anal Chem 61: 211
67. Torre M, Fernández-Gámez F, Lázaro F, Luque de Castro MD, Valcárcel M (1991) Analyst 116: 81
68. Chen D, Luque de Castro MD, Valcárcel M (1990) Talanta, 37: 1049
69. Chen D, Luque de Castro MD, Valcárcel M (1990) Anal Chim Acta 234: 345
70. Russell DA, Narayanaswamy R (1989) Anal Chim Acta 220: 75
71. Chen D, Luque de Castro MD, Valcárcel M (1992) Anal Chim Acta (in press)

16. Fluorescence Spectroscopy in Environmental and Hydrological Sciences

Marvin C. Goldberg and Eugene R. Weiner*

U.S. Geological Survey, P.O. Box 25046 MS 424, Denver Federal Center, Lakewood, CO 80225, USA

1 Introduction

One of man's earliest scientific endeavours involved observations of fluorescence and phosphorescence effects which, over time, led to our present understanding of the underlying electron structure of matter. In the newer disciplines such as hydrology and environmental science the earliest application of fluorescence involved tracing the flow of surface-waters from point to point. This rather minimal use of the fluorescence effect was the only early major hydrologic application and even today remains a workhorse method. In the past, a drawback to the use of fluorescence data to produce more detailed results was the lack of specificity. This has not been true for the past fifteen years. Advances in light sources, light detection, electronics, and optics have resulted in techniques for adding more specificity to fluorescence data. In much of the work discussed in this paper, excitation-emission matrix (EEM) patterns are utilized as pattern recognition techniques and as semi-quantitative techniques to follow the transport of natural and anthropogenic materials in hydrologic systems.

Analytical methods based on fluorescence emission are attractive because of their very high detection sensitivity. This asset has been mitigated by a lack of specificity, due to the fact that most room temperature fluorescence spectra have fairly broad emission peaks that are not easily distinguished one from another regardless of differences in the fluorophoric composition. Specificity can be improved by increasing the dimensionality of fluorescence measurements. Fluorescence spectroscopy is inherently more selective than absorption spectroscopy because at least five independent fluorescence variables can be measured which are characteristic of a given sample component:
- emission intensity as a function of excitation wavelength.
- emission intensity as a function of emission wavelength.
- excited state lifetime.
- emission polarization.
- quantum yield.

A common way to increase measurement specificity is by measuring the excitation spectrum as well as the emission spectrum of a fluorophoric substance. Even so, the specificity is still rather limited. Specificity can be further increased by collecting the entire excitation emission matrix (EEM) and presenting it in a single figure [1]. More specificity is obtained by employing polarized light measurements, such as the tangent delta, which determine the fluorescence depolarization attributed to both internal and molecular rotation of fluorophores in a polarized light field. Including the excited state lifetime of the chromophore adds additional specificity. Lakowicz [2] treats the theory of these additional dimensions of fluorescence measurements in detail. Warner et al. [3] have discussed the multiparameter nature of fluorescence measurements in detail and list fourteen different parameters that can be combined in various combinations for simultaneous measurement, thereby maximizing fluorescence selectivity with multidimensional measurements.

With the development of multidimensional techniques, fluorescence measurements are now included among the most useful methods of chemical analysis available to researchers [1, 4, 5]. This report describes some of the many ways in which fluorescence measurements are being used for studies of the environment with particular emphasis on water quality and hydrological applications of the EEM technique.

2 EEM Studies, Multi-Component Analysis, Spectral Signatures, and "Fingerprinting"

One of the most popular and visually satisfying methods of multidimensional analysis is to measure the EEM, which allows plotting the emission intensity at all combinations of excitation and emission wavelengths in a single three-dimensional graph, either as a contour diagram or as a topological surface [1, 3, 6, 7]. An EEM spectrum includes all the details of the excitation and emission fluorescence in a single figure. The measurement is usually made by holding the excitation wavelength constant and scanning the emission wavelength over the region of interest. Repeating this process at progressively higher excitation wavelengths yields spectral data that may be presented as a three dimensional surface, where the x-axis is the emission wavelength, the y-axis is the excitation wavelength, and the z-axis is the intensity of fluorescence emission.

The analysis of oil in water [8–10] was one of the first fingerprinting applications of EEM. More recently, Dudelzak, et al. [11] developed a catalog of 24 EEM spectral signatures from the Gulf of Finland and the Baltic Sea that allowed them to detect low levels of oil pollution (a few microliters/liter) in natural sea water containing dissolved organic matter (DOM) and to remotely identify groups of pollutants.

Using EEM, Coble et al. [12] detected at least three different fluorophores in Black Sea and measured changes in their relative abundances with depth. They concluded that EEM could distinguish sea water DOM of different types and sources. Zung et al. [13] used EEM to fingerprint algae and were able to detect spectral changes due to pollutant uptake by algae. Natural ocean samples containing algae were collected offshore from Savannah, Georgia, preconcentrated, and exposed to a series of substituted nitroaromatics, a pollutant known to disrupt the photosynthetic pathway. Quenching effects induced by the pollutants show that the toxicity of nitroaromatics depends on the ring substituent and, in addition, indicate some of the steps in the toxicity mechanism.

Ho et al. [14] successfully analyzed six component polynuclear aromatic hydrocarbon solutions by mathematical treatment of EEM data.

When in situ measurements are not required, certain aqueous pollutants can be analyzed by freezing them in a cryogenic solvent matrix, a procedure that offers large gains in spectral resolution (Shpol'skii effect) and makes spectral fingerprints more distinctive. Ariese et al. [15], showed that polycyclic aromatic hydrocarbons in complex environmental samples can be detected with great sensitivity by this approach. When used with tunable laser excitation, the Shpol'skii effect allows the spectrum of one compound to be isolated from others with great resolution.

The geologic stages of maturation of shale and coal samples can be identified by EEM in combination with synchronous scan (SS) spectra [16]. The authors suggest using the fluorescence patterns of oils and condensates as a screening tool to investigate compositional similarities among reservoired hydrocarbons.

Fluorescence lifetime selectivity has been incorporated into EEM measurements by McGown and coworkers [17–20], who compared phase-resolved EEM spectra (PREEM) collected at three different modulation frequencies with normal steady state EEM spectra. Their PREEM measurements of a binary system of benzo(k)fluoranthene and benzo(b)fluoranthene showed greater sensitivity due to the ability of the method to suppress scattered light and to selectively enhance spectral features of the individual components.

In a report on derivatizing reagents useful for HPLC analysis, Nelson et al. [21], thoroughly discuss the value of HPLC with EEM fluorescence detection for analyzing low levels of fluorescent compounds introduced into groundwaters for tracing water movement.

3 Non-EEM Fluorescence Methods

Surface and ground waters have been monitored with a wide variety of techniques. Fluorescent dyes are commonly used as tracers to follow water and pollutant movement [22–24]. Particle movement in surface waters has been tracked using model particles, either naturally fluorescent or made to be fluorescent by a variety of techniques [25–29]. Fiber optic techniques are being developed to allow remote sensing of fluorescent groundwater tracers [30, 31].

The use of fiber optics and UV laser excitation for the remote sensing of fluorescent groundwater contaminants in situ has been discussed by Chudyk et al. [32, 33]. Non-fluorescent contaminants can be detected with fluorescence optrodes, optical fibers that incorporate a reagent phase on the sensing end. The reagent phase reacts with an analyte to generate a fluorescence-based signal [34–36]. Kenny and coworkers [37, 38] have used a remote sensing field instrument based on fiber optics and UV excitation for measuring several important groundwater pollutants with high sensitivity.

There are many examples of using the natural (solar induced) fluorescence of natural waters as a measure of the amount of planktonic organisms and dissolved organic matter [39–46].

Natural fluorescence was used to distinguish between coastal plain and Piedmont river waters where they flow into nearshore areas of Georgia and North Carolina [47] in order to define circulation patterns in the nearshore area. In another case, natural fluorescence proved useful for monitoring agricultural pollutants originating from feedlots [48].

Natural fluorescence measurements were used in conjunction with other analytical techniques to fingerprint humic materials from several Venezuelan Blackwater Rivers [49]. Differences in fluorescence spectra were related to different vegetative and humification conditions on the individual rivers.

Traditional excitation and emission fluorescence spectroscopy, in combination with other spectroscopic methods, was applied to the study of Black Trona Water, a fossil water from the Green River Formation of Wyoming [50]. The fluorescence spectra indicated the presence of condensed aromatic hydrocarbons somewhat similar to those found in some petroleum fractions, components not observed by the other spectroscopic methods.

Synchronous scanning is probably the most prevalent non-EEM multi-dimensional technique currently in use. Synchronous scanning [7, 51, 52] is a simplification of the EEM technique in which both the excitation and emission wavelengths are scanned simultaneously with a typical offset of 50–100 nm. This approach generates a single two-dimensional slice through an EEM surface [53] and is especially useful for rapid screening analyses of complex mixtures.

Miano and coworkers [54] conducted the first systematic investigation of the fluorescence spectra of humic and fulvic acids of diverse origin, comparing excitation, emission, and synchronous scan (SS) spectra of samples extracted from soil sediments and natural waters. They found that excitation and synchronous scan spectra could be used to distinguish between humic and fulvic acids and among humic acids from different sources. They continued their investigations [55] by comparing the fluorescence spectra of several model humic acid-type polymers with natural humic substances to identify significant structural similarities. Their measurements confirmed the prominent role of carboxyl groups as substituents and indicated the extent of polycondensed and aromatic ring systems in natural humic substances.

Baudot et al. [56], did a comprehensive study of about 50 different polyaromatic hydrocarbons (PAH) by synchronous scan fluorescence spectrometry to evaluate the potential for creating a fingerprint library for these compounds and for correlating spectral features with molecular structural features. They were able to make many positive identifications of individual PAH compounds in mixtures and conclude that the technique has considerable potential for routine environmental analysis. In their measurements on an environmental sample, the SS results correlated closely with HPLC results.

Stainken and Frank [57] give a detailed procedure for using synchronous scan fluorescence spectroscopy for direct measurement of fluorescing industrial effluents and rhodamine dye tracer in river and ocean surf water samples. They were able to achieve fingerprint sensitivities in the part per trillion range.

Inman and Winefordner [58] showed that synchronous scan spectroscopy of mixtures of polynuclear aromatic hydrocarbons using a constant energy difference between the excitation and emission spectrometers yielded greater selectivity than when a constant wavelength difference was used. The authors offer guidelines for choosing the best values for the constant energy scanning increment.

Constant energy synchronous scan spectrometry at room and cryogenic temperatures was used to fingerprint gasoline and crude oil samples [59]. The authors conclude that the method is a simple, inexpensive and reliable way to identify environmental samples from different origins.

Shotyk and Sposito [60] reported that excitation and synchronous scan spectra could discriminate with respect to origin and structural differences among the important nonhumic organic materials found in the water soluble fraction of forest leaf litter. Organic acids derived from leaf litter are believed to be important weathering agents in forest soils, affecting soil acidity and mobilizing soil aluminium by complexation.

A variation on the usual form of synchronous scan fluorescence spectroscopy is *excitation resolved synchronous scan spectroscopy* (ERSF), in which a narrow excitation bandwidth is used in conjunction with a wide emission bandwidth. ERSF combines high resolution of the excitation spectral structure of a compound with high analytical sensitivity. Taylor and Patterson [61] have catalogued ERSF spectra of many aromatic organic compounds found in oils and shown that the method can be applied to the classification of different oils and to the study of oil degradation in soils.

The Shpol'skii effect at low temperatures has been used with the synchronous scan method [62] to further increase selectivity when analyzing polyaromatic hydrocarbons. At 77 K, constant energy synchronous scans could more readily be used for analyzing complex mixtures than at room temperature.

Vodacek and Philpott [63] investigated the interactions between dissolved organic carbon and trace metals through fluorescence quenching effects, to assess the applicability of airborne oceanographic lidar to the detection of trace metals in marine waters. Philpott and Vodacek [64] found that the dissolved organic carbon (DOC) content of lake water from 49 lakes could be correlated closely with an empirical factor derived from several experimental quantities, including the integrated laser-induced fluorescence of DOM, the Raman water signal and a spectral bandwidth quantity. Vodacek [65] reviews the use of SS spectroscopy for fingerprinting DOM in surface waters and gives useful details regarding the assignment of certain SS peaks in humic substances to specific molecular precursors and their degradation products. He also discusses the application of the SS method to airborne remote sensing. The influence of pH on the fluorescence of dissolved organic matter in estuarial waters was examined by Laane [66], who developed a formula to express fluorescence at a fixed pH.

The influence of binding between hydrophobic organic pollutants and naturally occurring organic colloids, such as humic materials, on the transport of pollutants in groundwater was investigated by a fluorescent quenching technique [67]. The basis of the method is the quenching of polycyclic aromatic hydrocarbon fluorescence upon association with colloidal organic carbon materials. Partition coefficients between the hydrophobic pollutants and organic colloids and transport retardation factors were determined. Limits of the approach were discussed in detail.

The interactions of humic materials with pollutants in water has been studied with fluorescence polarization [68, 69]. These workers studied the binding of metal ions to humic substances in natural waters by the quenching of the natural fluorescence of fulvic acid by bound heavy metal ions and the binding of organic fluorophores to fulvic acid by rotational depolarization. The change in polarization of perylene fluorescence when perylene binds to fulvic acid allowed the determination of a binding constant. Seitz [68] also obtained conformational information about a soil fulvic acid by rotational depolarization, concluding that the soil fulvic acid has a flat extended shape.

4 Determination of Humic Acid Dissociation Constants

Fluorescence excitation spectra obtained at different pH values can be used to determine humic acid dissociation constants [70]. Reliable values for humic acid dissociation constants cannot be found by direct measurements because of the heterogeneity of humic materials, the variety of experimental isolation procedures, and the lack of a specific structural formula. Isosbestic points in pH dependent fluorescence excitation spectra indicate an equilibrium between singly and doubly deprotonated forms of humic acid and allow a calculation of the second ionization constant. An equation was derived that relates emission intensities, pH, and the second ionization constant, K_{a2}. The value obtained from measurements on a soil humic acid from Beaver Hills, Alberta was $K_{a2} = 3.1 \times 10^{-9}$, in the same range as values based on other indirect techniques on different humic acid samples. Thus, the fluorescence isosbestic point method offers an additional tool for a needed measurement that is beset with inherent uncertainties.

5 Physical Characteristics and Fluorophore Composition of Fulvic and Humic Acids

Fluorescence depolarization and excited state lifetime measurements were used in conjunction with EEM spectra to characterize several physical and chemical parameters of fulvic and humic acids. Since different fluorophores in a mixture will in general have different lifetimes, lifetime measurements can increase the specificity of fluorescence analysis [1, 71–73]. Under the proper conditions, depolarization measurements can indicate the rate of rotation of fluorophores in solution and allow estimates of molecular size and shape [2, 74, 75].

Goldberg and coworkers [74, 75] report on rotational depolarization measurements of both fulvic and humic acids obtained from natural waters, finding equivalent sphere volumes for fulvic acids that are smaller than found by Seitz [68]. They used dynamic depolarization techniques to obtain conformational information, finding that the water fulvic acid was spherical in the pH range 3–11, changing shape at higher and lower pH values. Using phase modulated fluorometry, Goldberg and Negomir [75] concluded that there are two or more different fluorophores present in their fulvic acid.

6 Hydrologic Transport of Phenol and *p*-Cresol in the South Platte River Colorado

We have applied the EEM technique to the determination of movement of water-borne materials in the South Platte River, Colorado. The Platte River rises in the mountains of Colorado descending from a height of 4.3 km above sea level and flows through a rather uninhabited area to a plain which is about 1.5 km above sea level. As the river emerges from the mountains, it encounters increasing urban density at the outskirts and boundaries of the city of Denver, Colorado, U.S.A. As it exits from Denver, it then flows on to the next city downstream. Water samples were taken along a 42 km stretch of this watercourse. Samples were taken in a downstream direction at 0 km, (Chatfield Reservoir immediately upstream from Denver), 8.5 km (Belleview St.), 15.7 km (Exposition Avenue), 25 km (8th Avenue), 28.5 km (58th Avenue), 36.8 km (72nd Avenue) and 41.3 km (88th Avenue). Two materials that are indicative of the waste products of urban populations are *p*-cresol and indole. *p*-cresol is used as an additive to household washing formulations and is found in soaps and other domestic cleaning powders. Indole is an amino acid produced in the small intestine of man and other animals. Both of these materials are delivered to the watercourse at various entry points through drains, surface runoff, and river tributaries both above ground and below ground.

7 Gas Chromatographic Measurements

Indole and *p*-cresol were independently measured by gas chromatography at these sampling points both in the sediments and the water (see Figs. 16.1, 16.2). Sediments extracted with a soxhlet extractor with methylene dichloride and hexane according to the procedure found in Goldberg [76]. They were concentrated in a Kuderna–Danish evaporator and then analyzed by gas chromatography. The water samples were extracted into hexane and methylene dichloride and similarly analyzed by gas chromatography. The values found,

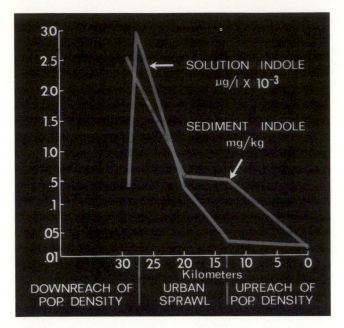

Fig. 16.1. Concentrations of indole in sediment and water at selected sampling points in the South Platte River, measured by gas chromatography

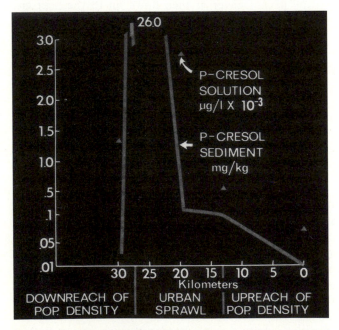

Fig. 16.2. Concentrations of *p*-cresol in sediment and water at selected sampling points in the South Platte River, measured by gas chromatography

both in the sediments and the water are illustrated in Figs. 16.1 and 16.2. No measurable indole was found at the 0 km point, which is the low mountain areas and far above the urban sprawl of Denver. There is a small initial amount of p-cresol, at the Chatfield Reservoir sampling point; this is attributed to leaching of wood found in the mountain forests. As the river flows toward the city of Denver, the concentrations of indole vary from almost zero concentration at 0 km upreach of the population density, to a high value at 72nd Avenue, which is just before the sewage treatment plant and about 90% through the area of urban density. They then decrease as the river flows past the sewage treatment plant (see Figs. 16.1 and 16.2). The concentrations of p-cresol in water, start at low levels and increase to a maximum value at 72nd Ave. This trend parallels that of indole but p-cresol concentrations start at a higher level and end at a higher level. The sediments also contain p-cresol and indole and load these materials in relatively similar fashion to the concentrations in water. The concentrations in the sediments are much larger than those found in water indicating that the sediments act as sinks that are in equilibrium with the water concentrations. The p-cresol and indole in the sediments show the same concentration trends as found in the water.

8 Fluorescence Measurements and Results

A similar study was conducted in which samples taken at the same Platte River locations were analyzed for p-cresol and indole using EEM fluorescence patterns. Several precautions are important to observe when working with environmental samples. It is necessary to limit the self absorption and inner filter effects as much as possible, but these are easy to avoid simply by using a dilute solution for the fluorescence measurements. For most work a sample diluted to an optical density of 0.05 is sufficient. Also, it is possible that environmental samples will contain interfering fluorophores other than those being monitored or contain ions that quench fluorescence. In most instances where the methods described in this paper are applied it is necessary to have additional analytical information concerning the composition of the sample to avoid errors that could be caused by interfering fluorophores or quenching agents. A complete analysis of a few samples in the aquatic system being measured will serve to determine these factors and allow the correct application of fluorescence measurement techniques to many additional samples taken from the same watercourse. The EEM matrix will show the absorption and emission data for the intense fluorophores and the wavelength separation between fluorophore peaks can be easily observed.

Rhodamine B is a good internal standard for EEM natural water measurements. Its excitation and emission maxima are distinct from that of most materials found in natural settings. We have used the peak height of rhodamine-B dye as an internal standard to normalize each EEM spectrum. Once this is done, addition and subtraction of complete EEM spectra is possible. This procedure also was used to make correlations between a standard, obtained by measuring the fluorescence emission intensity from a known compound in pure solution, and the fluorescence emission intensity of that material in an environmental sample. Where pattern recognition is the key usage, subtraction of EEM spectra is facilitated by normalization to rhodamine-B as the internal standard.

9 Pattern Recognition

Humans have excellent visual pattern recognition skills because it is instinctive to be able to recognize danger and distinguish between safe and unsafe surroundings. Fluorescence EEM spectra can be viewed as topological surfaces or as contours (see Fig. 16.3) and the same visual recognition skills that were developed for survival can be applied to studying these patterns.

Figure 16.3a is a contour diagram of the fluorescence pattern found in samples from Chatfield reservoir. The Chatfield EEM pattern characterizes the water that drains into Chatfield reservoir from the surrounding watershed. In Figs 16.3a,b,c,d, the contour pattern changes can be seen that result from materials entering the river below Chatfield reservoir. These changes are particularly evident in the high energy region of the EEM spectra where indole and p-cresol appear, and qualitatively parallel the measurements of these pollutants that were earlier determined by gas chromatography. In examining these figures, one should note changes in overall pattern and peak shapes rather than absolute fluorescence intensity. The patterns corresponding to the Chatfield reservoir and Belleview samples are almost identical, indicating the organic loads at these locations to be quite similar. At 8th Avenue the high energy portion (200–260 nm excitation, 250–400 nm emission) shows additional structure, indicating that new fluorophores have entered the river. At 72nd Avenue the pattern (not shown) has additional changes from that found at Belleview, indicating that further organic load additions and losses occurred between Belleview and 72nd Avenue. At 88th Avenue the EEM pattern shows the continued presence of large amounts of fluorescent material probably due to discharge from the sewage treatment plant. A base organic load in the water continues to be present and is transported out of the urban area below 72nd Avenue.

In Figures 16.4a,b,c,d (8th Avenue minus Chatfield), (72nd Avenue minus Chatfield), (88th minus Chatfield) and (72nd Avenue minus Chatfield minus standard indole spectrum) the contour spectra have had the Chatfield fluorophore contribution subtracted. The resultant spectra represent only the fluorophores added to the river downstream of Chatfield Reservoir. The increase and decrease of the high-energy fluorophores, indole and p-cresol, is evident. To confirm this figure 16.4d shows the spectrum of 72nd Avenue minus Chatfield minus indole, which has almost no high-energy fluorescence emission. In fact this spectrum has almost no fluorescence signal throughout the entire wavelength range. The EEM spectra clearly show less indole and p-cresol at 88th Avenue than at 72nd Avenue. These results are qualitatively the same as the gas chromatographic analyses.

Both the EEM and gas chromatographic data lead to the same conclusions, that the amount of p-cresol and indole in the Platte River greatly increases between 8th Avenue and 72nd Avenue then rapidly decreases downstream.

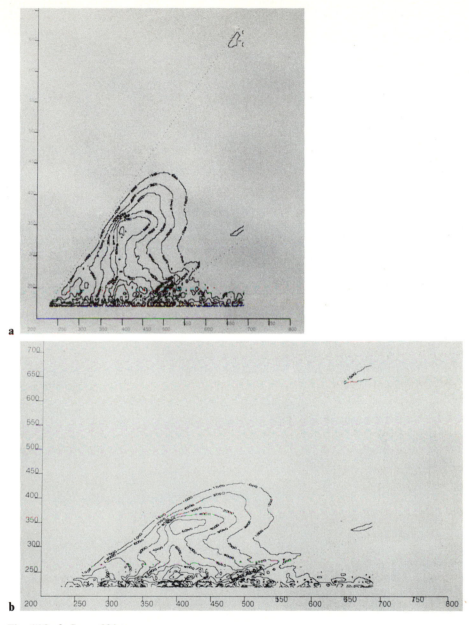

Fig. 16.3.a,b. See p. 224

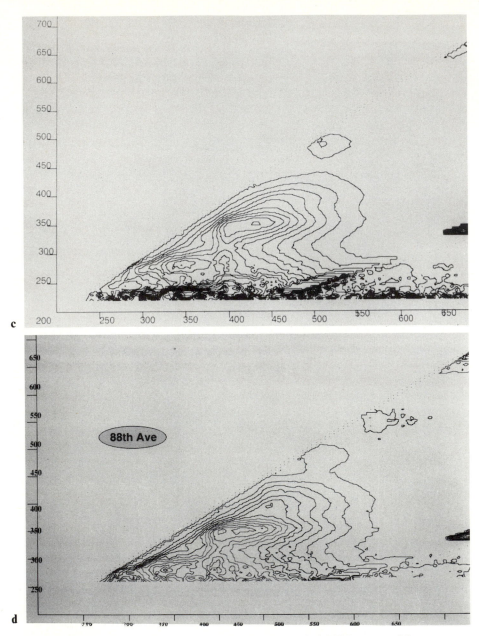

Fig. 16.3. a EEM contour of fluorescence from Chatfield Reservoir; **b** EEM contour of fluorescence from Bellview; **c** EEM contour of fluorescence from 8th Ave.; **d** EEM contour of fluorescence from 88th Ave.

10 Semi-Quantitative Method for Calculation of Known Materials in EEM Spectra

An internal standard of rhodamine-B is added in the same amount to each sample. The optical density of each sample is measured at the excitation–emission peak maximum. This maximum point is chosen by examination of the

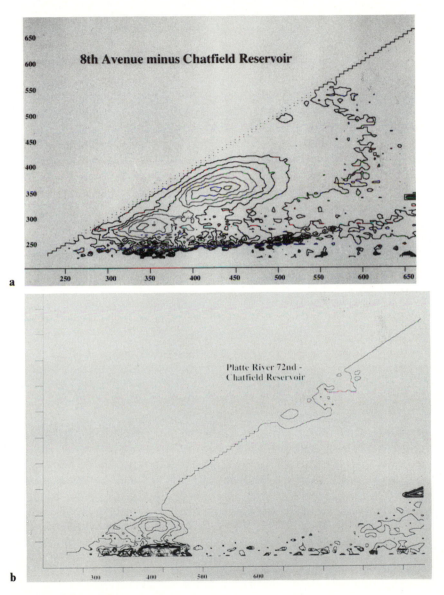

Fig. 16.4a,b. See p. 226

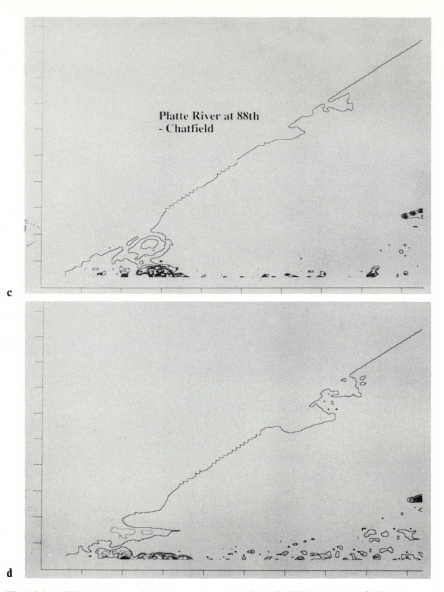

Fig. 16.4. a EEM contour of 8th Ave. minus Chatfield; **b** EEM contour of 72nd Ave. mınus Chatfield; **c** EEM contour of 88th Ave. minus chatfield; **d** EEM contour of 72nd Ave. mınus Chatfield minus indole

EEM spectrum and if several maxima are available, the one which is more isolated from the other fluorophoric peaks is used. A standard of the material being measured in the environment is run at the same emission and excitation settings (see Fig. 16.5). The fluorescence intensity at an emission, excitation peak maximum is used to determine the amount of the material in the environmental

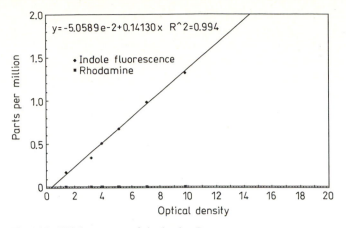

Fig. 16.5. EEM spectrum of rhodamine B

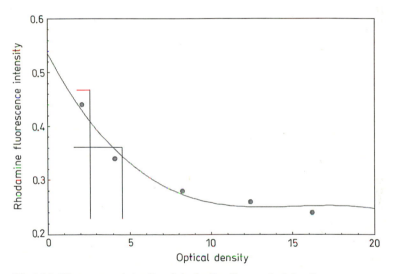

Fig. 16.6. Fluorescence intensity of rhodamine B vs. optical density

sample but before that can be done corrections must be made for changes in the fluorescence intensity of the internal standard, rhodamine-B, due to differences in optical density. These differences are normally found in different samples and the standard (see Fig. 16.6). As seen in Fig. 16.6 the fluorescence intensity of rhodamine-B changes with increasing optical density of solution. A standard curve is obtained for each material that is to be quantified in the EEM spectrum to determine the effect that the change in optical density has on the fluorescence intensity of the internal standard (rhodamine-B). When the correction factor is obtained, it is used to adjust the fluorescence intensity of rhodamine-B from the measured value and this corrected value is used in the calculations that normalize all the spectra, both samples and standards.

Once the corrected fluorescence intensity of the sample is determined, the intensity values can be normalized for each EEM spectrum and the normalized, corrected fluorescence intensity is compared with the normalized fluorescence intensity of the standard to obtain a semi-quantitative measurement of the known fluorophore. All corrections are made at the emission and excitation maxima of the fluorophores.

Because there can be other absorbers and emitters at these wavelengths, the data are only semi-quantitative. Successful applications of this procedure to environmental samples requires qualitative information about the identity of the materials that compose the samples. Additional analyses by other analytical techniques are required. These other analyses are used to identify the materials of interest in the watercourse. After a given material is known to exist in a watercourse, and if it is a relatively strong fluorescing component, its transport properties can be followed by EEM fluorescence spectra taken at various locations. This technique can be applied with the expectation that changes in

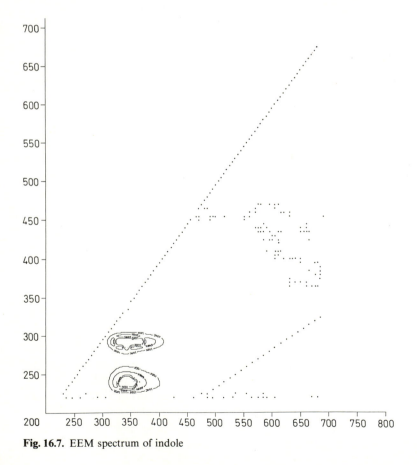

Fig. 16.7. EEM spectrum of indole

concentration occurring along a watercourse can be monitored with sufficient accuracy to pinpoint the changes in concentration and the locations of input and output from the hydrologic system.

The EEM spectra of indole (Fig. 16.7) and p-cresol (see Fig. 16.8) clearly show areas of peak excitation and emission intensity which are distinct one from another. These spectra were used to choose the excitation and emission values that were used in calibrating the fluorescence intensity of each material to the concentration of that material.

The South Platte River EEM spectra were re-examined using the semi-quantitative methods described above and the results were compared with the gas chromatographic analyses. The fluorescence results are seen in Fig. 16.9. At Chatfield reservoir, which is the 0 km point in the watercourse, indole concentrations are less than 0.3 ppm in water. This increases by a small amount about 8.5 km further downstream, at the Belleview location however, the values are slightly below 0.4 ppm. At exposition, an additional 7.2 km downstream the value is about 0.8 ppm and increases to 1.2 ppm at 58th Avenue which is 28 km

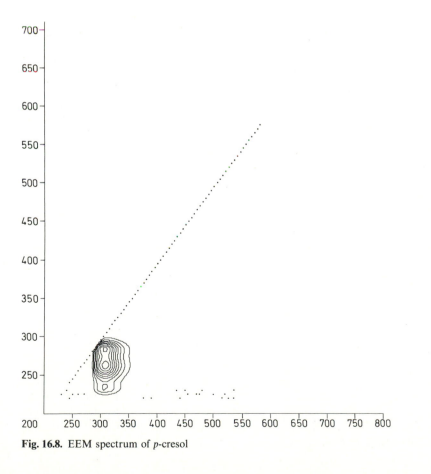

Fig. 16.8. EEM spectrum of p-cresol

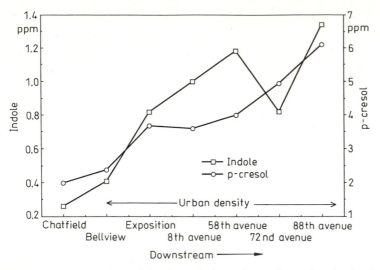

Fig. 16.9. Concentrations of indole at, designated sampling points in the South Platte River measured by fluorescence spectroscopy

downstream. Indole being a sewage product is reduced somewhat at the sewage treatment plant, located at the 72nd Avenue sample point, but increases to almost 1.4 ppm at 88th Avenue. The area of higher urban density through which the river flows increases between Belleview and 88th Ave which is 8 to 43 km downstream from the 0 km position. The data on indole concentrations, although not exactly the same as that from the gas chromatographic study, show the same general trends and pinpoint the areas from 58th Avenue to 88th Avenue as having high input amounts of indole compared to that found at Chatfield reservoir, the 0 km point.

A similar examination of the data for *p*-cresol, (see Fig. 16.9) shows a value of 2.0 ppm at the 0 km point. The concentration increases to 4 ppm at exposition, 15.6 km downstream, and then steadily increases up to 6 ppm at 88th Ave which is 43.2 km downstream. *p*-cresol is not easily removed from water by primary and secondary sewage treatment nor is it sorbed completely on the sediment, thus *p*-cresol continues to build up in concentration in the river as urban sources continue to provide this chemical to the aquatic environment.

Humic materials can also be semi-quantitatively estimated. An analysis of the dissolved organic material (DOM) as standardized against fulvic acid was made. This DOM is normally considered to show up at excitation wavelength of 364 nm and emission wavelength of 445 nm. Figure 16.10 shows very little DOM at Chatfield Reservoir but values up to 27 ppm are found at Belleview. The DOM drops as the river flows through the next 20 km then rises to a high value of 44 ppm at 72nd Avenue. This location receives the output of the sewage treatment plant. A large amount of DOM continues to be present at 88th Ave which is 43 km downstream from the 0 km point. At this location the river has traversed the high urban density area.

Discussion. Indole and *p*-cresol have parallel buildup in the South Platte River. Both being at very low levels prior to the rivers entering the urban density area, then building up to values 10 times greater in amount for indole and 3 times greater in amount for *p*-cresol as the river flows through a densely populated urban area. DOM calibrated as fulvic acid equivalents, however, (see Fig. 16.11) does not continuously increase but is 3 times greater at the 41 km point downstream than at the 0 point in Chatfield Reservoir.

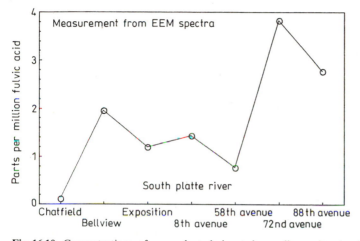

Fig. 16.10. Concentrations of *p*-cresol at designated sampling points in the South Plate River measured by fluorescence spectroscopy

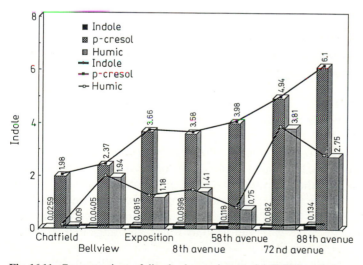

Fig. 16.11. Concentrations of dissolved organic material (DOM) at designated sampling points in the South Platte River measured by fluorescence spectroscopy

There are two conclusions to this work. First urban areas add large amounts of domestic chemicals to watercourses that flow through them and second, the EEM fluorescence spectroscopy technique applied here is capable of indicating these changes in content and transport. The fluorescence technique can be usefully employed where trends are more important than exact analytical results. It should be noted that the cost of making the analysis and the time spent in making the analysis are much less with the fluorescence methods than with conventional GC/MS or LC/MS methods.

11 EEM Spectra of Humic Fractions

Humic acid was fractionated on the basis of it hydrophobicity and proton affinity resulting in six fractions. These fractions are easily distinguished by EEM spectra in the topologic surface presentation mode. These materials were sized by fluorescence depolarization measurements (75). The sizes ranged from 9800 to 1300 equivalent spherical molar volume in cubic centimeters. Presented in order of size the largest fraction, the hydrophobic humic strong acid is seen in Fig. 16.12a. It has three distinct peaks and the excitation and emission maxima of each peak can be clearly identified. Similarly, the hydrophilic acid (Fig. 16.12b), the hydrophobic humic weak acid (Fig. 16.12c), the hydrophobic strong acid (Fig. 16.12d), the hydrophobic weak acid (Fig. 16.12e) and the hydrophobic neutral (Fig. 16.12f) show differences of two and three distinct peaks. In some cases the lower energy peak is larger than the higher energy peak, and in other cases this trend is reversed. It is possible to uniquely identify each material from its EEM spectrum. The hydrophobic strong acid is also known as fulvic acid. It has been established that in most natural surface waters the predominant form of humic material is fulvic acid. Comparison of the fulvic acid EEM spectrum with water from the Suwannee River Georgia, showed identical EEM spectra. It was concluded that fulvic acid is the predominant fluorescent material in the Suwannee river and that the humic materials in the river are in the form of fulvic acid (Fig. 16.13).

12 Groundwater Pollution Mechanisms

EEM fluorescence spectroscopy was used to identify an indirect mechanism of ground water pollution. Chlorinated organic solvents, primarily trichloroethylene (TCE), injected into the unsaturated zone of an aquifer, displaced soil pore waters and caused them to enter the mobile ground water. Highly colored dissolved organic carbon (DOC) materials, hitherto immobilized in the pore water, became free to move with the ground water and appeared in down-gradient wells.

Trichloroethylene was used as a degreaser from 1973 to 1985 at Building 24 at Picatinny Arsenal (see Fig. 16.14) from 1973 to 1985. [77] The TCE was

Okeefenokee Strong Hydrophobic
Humic Acid

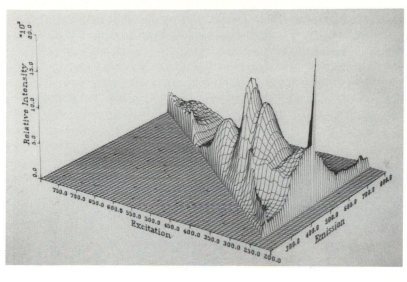

a

Okeefenokee Hydrophilic
Acid

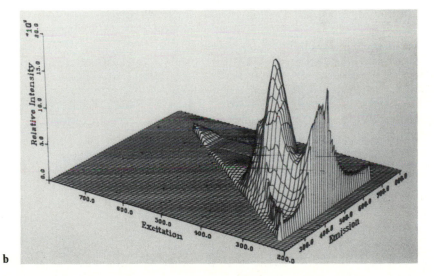

b

Fig. 16.12a,b. See p. 235

Okeefenokee Weak Hydrophobic
Humic Acid

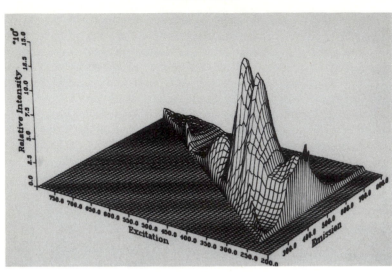

c

Okeefenokee Strong
Hydrophobic Acid

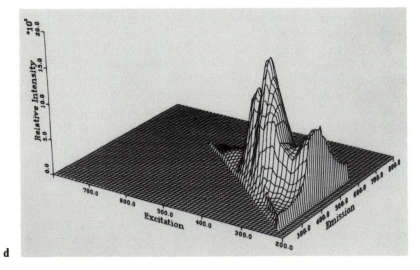

d

Fig. 16.12c,d. See p. 235

Okeefenokee Weak
Hydrophobic Acid

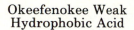

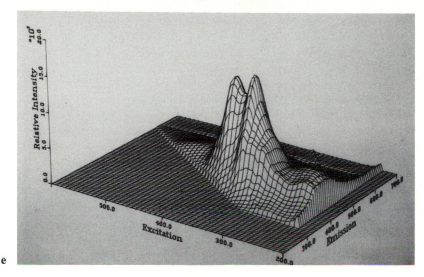

e

Okeefenokee Hydrophobic
Neutral

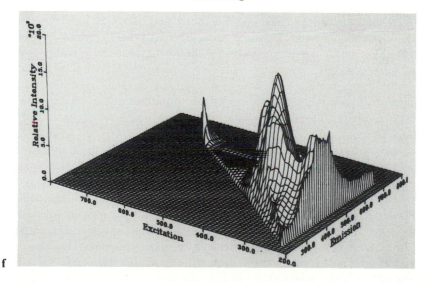

f

Fig. 16.12. a EEM spectrum of hydrophobic humic strong acid fraction; **b** EEM spectrum of hydrophilic acid fraction; **c** EEM spectrum of hydrophobic humic weak acid fraction; **d** EEM spectrum of hydrophobic strong acid (fulvic acid) fraction; **e** EEM spectrum of hydrophobic weak acid fraction; **f** EEM spectrum of hydrophobic neutral fraction

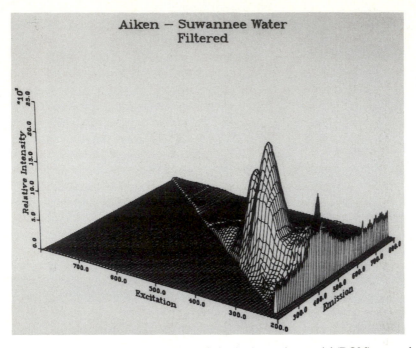

Fig. 16.13. Comparative concentrations of dissolved organic material (DOM), *p*-cresol and indole at designated sampling points in the South Platte River, Colorado, measured by fluorescence spectroscopy

discharged from an overflow pipe into a 1 meter deep dry well which created a TCE plume that contaminated ground water and extended from building 24 to Green Brook Pond. In the groundwater between these two points, water samples were colored brown and had dissolved organic matter analysis of 12 mg/l (milligrams per liter) which is relatively high. In other locations beyond this plume, ground waters were also sampled and found not to contain brown color. Samples were collected at well 9E which is in the plume of the TCE flow, and an EEM spectrum was taken which is shown in Fig. 16.15. The characteristic pattern of humic materials is noted with peak excitation in the 360 nm range and peak emission near 450 nm. Because of noise in the sample, peaks at excitation between 200 and 260 are not distinguishable. Figure 16.16 is the spectrum of water from a deep well 90 meters upgradient from the solvent dumping site. It collects water from the same aquifer as the downgradient well but shows no observable fluorescence. Figure 16.17 shows the EEM contour spectrum of a sample of ground water taken in a lysimeter. The lysimeter extracts the pore water in the soil column. It is readily seen that the EEM spectrum of the pore water (Fig. 16.17) and the EEM spectrum from the contaminated plume (Fig. 16.15) are identical. The conclusions from this observation are that the TCE is able to displace the pore water thus mobilizing the dissolved organic material into the unsaturated zone. By comparison of the

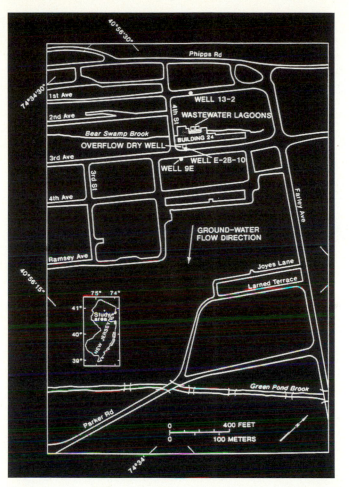

Fig. 16.14. Map of degreasing facilities and ground water TCE plume at picatinney arsenal, MD

EEM contour spectra taken of samples in the plume and in the pore water from the lysimeter, with those spectra of the humic fractions, it is seen that the pore water has a high concentration of dissolved organic matter with an EEM spectrum very similar to that of a humic material.

13 Conclusions

By increasing the dimensionality of fluorescence measurements they lend themselves in a more useful manner to the study of environmental and hydrological systems. The excitation, emission matrix spectrum is shown to be valuable as a tool to determine the changes in concentration of specific fluorescent compounds in a watercourse. In the South Platte River, Colorado, the influx of

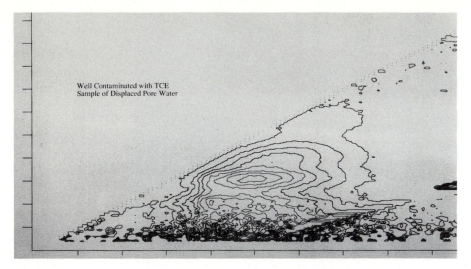

Fig. 16.15. EEM spectrum of a sample taken at well 9E, in the groundwater plume of the TCE contamination

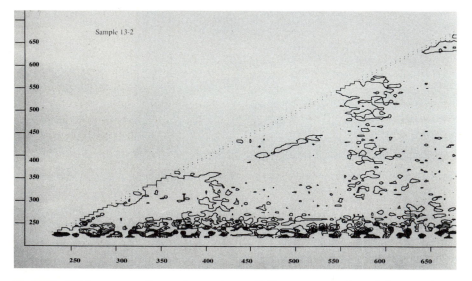

Fig. 16.16. EEM spectrum of water from a deep well 90 meters upgradient of the groundwater contamination plume

p-cresol and indole were determined by gas chromatographic analyses and by examination of the fluorescence EEM spectra. A semi-quantitative technique was applied to measure the amount of these materials at the same sampling points. It was established that the fluorescence analyses resulted in the same data trends and were close in absolute values to those obtained by GC analysis.

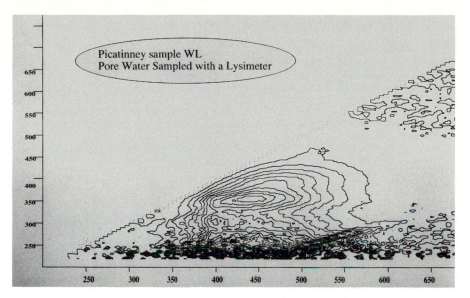

Fig. 16.17. EEM spectrum of pore water taken with a lysimeter, outside the TCE groundwater plume

EEM spectra were also taken of samples from Picatinny Arsenal in New Jersey. It was shown that samples in the plume of a TCE spill were very high in dissolved organic matter whereas samples taken in the same aquifer outside the TCE plume did not contain large amounts of dissolved organic matter. By relating the EEM spectra of 6 isolates of humic materials to the spectrum of the organic material it was possible to state that the organic compound displaced by the TCE was high in humic content. Samples from the pore water of this aquifer had the identical EEM spectrum to that found in the TCE plume. It was concluded that the humics that were stored in the pore water were mobilized by the TCE and were released to the ground water.

Humic fractions that were separated on the basis of their proton affinity and hydrophobicity were characterized with fluorescence EEM spectra. By matching these spectra with a sample of water taken from the Suwannee River, Georgia it was possible to recognize that the fluorescence from the river sample was due to fulvic acid. This fact is supported by finding that the most general form of aquatic humic material in river waters is fulvic acid.

The natural fluorescence of water samples can be used to observe the change in character of dissolved components in the sample and to monitor their entrance, egress and behavior while in the watercourse. EEM fluorescence patterns offer the means to qualitatively describe the transport of materials in hydrologic systems without resorting to specific molecular analysis. In many instances this provides the environmental scientist and hydrologist with data about changes in the system in a relatively short time frame, at low cost.

14 References

1. Weiner ER, Goldberg MC (1982) American Laboratory 14: 91
2. Lakowicz (1983) Principles of fluorescence spectroscopy, Plenum, New York
3. Warner IM, Patonay G, Thomas MP (1985) Anal Chem 57: 463A
4. Patonay G, Nelson G, Warner IM (1987) Prog Anal Spectrosc 10: 561
5. Goldberg MC, Weiner ER (1989) In: Goldberg MC (ed) Luminescence applications in biological, chemical, environmental, and hydrological sciences, ACS Symposium Series No. 383, American Chemical Society, Washington, DC, chap 1
6. McGown LB, Bright FV (1984) Anal Chem 56: 1400A
7. Warner IM, McGown LB (1982) CRC Critical Reviews in Analytical Chemistry 13: 155
8. Horning AW (1974) Marine Pollution Monitoring, NBS Special Publication 409, Symposium Proc., NBS, Gaithersburg, MD, May 1974
9. Keizer PD, Gordon DC, Jr., Dale J (1977) J Fish Res Board Can 34: 347
10. John P, Soutar I (1981) In: G.B. Crump (ed) Petroanalysis, John Wiley, London, p 106
11. Dudelzak AE, Babichenko SM, Poryvkina LV, Saar KJ (1991) Applied Optics 30: 453
12. Coble PG, Green SA, Blough NV, Gagosian RB (1990) Nature 348: 432
13. Zung JB, Woodlee RL, Ming-Ren SF, Warner IM (1990) Intern J Environ Chem 41: 149
14. Ho C-N, Christian GD, Davidson ER (1980) Anal Chem 53: 92
15. Ariese F, Gooijer C, Velthorst NH, Hofstraat JW (1991) Fresnius J Anal Chem 339: 722
16. von der Dick H, Kalkreuth W (1986) Organic Geochemistry 10: 633
17. Nithipatikom K, McGown LB (1987) Applied Spectroscopy 41: 1080
18. McGown LB, Millican DW (1988) Applied Spectroscopy 42: 1084
19. Millican DW, Nithipatikom K, McGown LB (1988) Spectrochimica Acta 43B: 629
20. Millican DW, McGown LB (1989) Anal Chem 61: 580
21. Nelson DA, Bush JH, Beckett JR, Lenz DM, Rowe DW (1989) In: Goldberg MC (ed) Luminescence applications in biological, chemical, environmental, and hydrological sciences, ACS Symposium Series No. 383, American Chemical Society, Washington, DC, chap 12
22. Smart PL, Laidlaw IMS (1977) Water Resour Res 13: 15
23. Morfis A, Paraskevopoulou P (1986) Proceedings of the 5th International Symposium on Underground Water Tracing, Athens (Contains many papers that deal with the use of fluorescent dyes as groundwater tracers)
24. Wilson Jr. JF, Cobb ED, Kilpatrick FA (1986) In: Fluorometric procedures for dye tracing, Techniques of Water-Resources Investigations of the United States Geological Survey, Book 3, Applications of Hydraulics, Revised 1986, TWI 3-A12, United States Government Printing Office, Washington, DC
25. Kirk RM, McLean RF, Burgess JS, Reay MB (1974) Geologiska Föreningens i Stockholm Förhandlingar, 96: 208
26. Duane DB, James WR (1980) J Sediment Petrol 50: 9
27. Herring JR (1981) In: Kavanaugh MC, Leckie JO (eds) Particulates in Water, Advances in Chemistry 189, American Chemical Society
28. Newman K, Morel, FMM, Stolzenbach KD (1990a) Environ Sci Technol 24: 506
29. Newman KA, Frankel SL, Stolzenbach KD (1990b) Environ Sci Technol 24: 513
30. Hirschfeld T, Deaton T, Milanovich F, Klainer S (1983) Opt Eng 22: 527
31. Hirschfeld T, Deaton T, Milanovich F, Klainer S, Fitzsimmons C (1984) Project Summary – Feasibility of using fiber optics for monitoring groundwater contaminants, U.S. Environmental Protection Agency, Environmental Monitoring Systems Lab, Las Vegas, NV, January 1984
32. Chudyk WA, Carrabba MM, Kenny JE (1985) Anal Chem 57: 1237
33. Chudyk W, Pohlig K, Exarhoulakos K, Holsinger J, Rico N (1990) In: Nielsen DM, Johnson AI (eds) Ground water and vadose zone monitoring, ASTM STP 1053, American Society for Testing and Materials, Philadelphia, PA, p 266
34. Seitz WR (1984) Anal Chem 56: 16A
35. Peterson JI, Fitzgerald RV, Buckhold DK (1984) Anal Chem 56: 62
36. Munkholm C, Walt DR, Milanovich FP, Klainer SM (1986) Anal Chem 58: 1427
37. Kenny JE, Jarvis GB, Chudyk WA, Pohlig KO (1987) Analytical Instrumentation, 16: 423
38. Kenny JE, Jarvis GB, Chudyk WA, Pohlig KO (1989) Ch. 14 In: Goldberg MC (ed) Luminescence applications in biological, chemical, environmental, and hydrological sciences, ACS Symposium Series No. 383, American Chemical Society, Washington, DC

39. Christman RF, Minear RA (1971) In: Faust SD, Hunter JV (eds) Organic compounds in aquatic environments, M. Dekker, New York, p 119
40. Egan WG (1974) Marine Technology Society Journal 8: 40
41. Smart PL, Finlayson BL, Rylands WD, Ball CM (1976) Water Res 10: 805
42. McMurray G (1978) In vivo fluorescence studies in the San Francisco Bay, U.S. Geological Survey Professional Paper No. 1100, U.S. Government Printing Office, Washington, DC p 162
43. Alpine AE, Cloern JE, Cole E (1979) EOS Transactions, paper no. O63, Amer Geophys Union, 60: 851
44. Broenkow WW, Lewitus AJ, Yarbrough MA, Krenz RT (1983) Nature, 302: 329
45. Lyutsarev SV, Gorshkova ON, Chubarov VV (1984) Oceanology 24: 71
46. Chen RF, Bada JL (1989) Paper No. 58, Abstracts of Papers, 197th ACS National Meeting, Dallas Texas, April 9–14, 1989, GEOC
47. Willey JD, Atkinson LP (1982) Estuarine, Coastal, and Shelf Science 14: 49
48. Lakshman G (1975) Water Resources Research 11: 705
49. Vegas-Vilarrúbia T, Paolini JE, Miragaya JG (1988) Biogeochemistry 6: 59
50. Branthaven JF, Barden JF (1983) Org Geochem 4: 117
51. Vo-Dinh T (1982) In: Eastwood D (ed) New directions in molecular luminescence, Subcommittee E13.06 on Molecular Luminescence: Philadelphia, p 5
52. Rubio S, Gomez-Hens A, Valcarcel M (1986) Talanta 33: 633
53. Weiner ER (1978) Anal Chem 50: 1583
54. Miano TM, Sposito G, Martin JP (1988) Soil Sci Soc Am J 52: 1016
55. Miano TM, Sposito G, Martin J (1990) Geoderma 47: 349
56. Baudot Ph, Viriot ML, André JC, Jezequel JY, Lafontaine M (1991) Analusis 19: 85
57. Stainken DM, Frank U (1990) In: Friedman D (ed) Waste testing and quality assurance, vol 2, ASTM STP 1062, American Society for Testing and Materials, Philadelphia, PA, p 381
58. Inman Jr. EL, Winefordner JD (1982a) Anal Chem 54: 2018
59. Files LA, Moore M, Kerkhoff MJ, Winefordner JD (1987) Microchemical Journal 35: 305
60. Shotyk W, Sposito G (1990) Soil Sci, Soc Am J 54: 1305
61. Taylor TA, Patterson HH (1987) Anal Chem 59: 2180
62. Inman Jr. EL, Winefordner JD (1982b) Analytica Chimica Acta 141: 241
63. Vodacek A, Philpot W (1983) Detection of dissolved organic substances and associated trace metals in water using laser fluorosensing, Proc 2nd Annual New York State Sympos on Atmospheric Decomposition, p 177
64. Philpot WD, Vodacek A (1989) Remote Sens Environ 29: 51
65. Vodacek A (1989) Remote Sens Environ 30: 239
66. Laane RWPM (1982) Marine Chemistry 11: 395
67. Backhus DA, Gschwend PM (1990) Environ Sci Technol 24: 1214
68. Seitz WR (1981) Trends in Analytical Chemistry 1: 79
69. Roemelt PM, Seitz WR (1982) Environ Sci Technol 16: 613
70. Goldberg MC, Cunningham KM, Weiner ER (1987) Can J Soil Sci 67: 715
71. Love LBC, Upton LM (1980) Anal Chem 52: 496
72. Demas JN (1983) Excited state lifetime measurements, Academic, New York, chap 1
73. Goldberg MC (1984) The fluorescence lifetimes of fulvic acid chromophores, 26th Rocky Mountain Conference, August 5–9, 1984, Denver, CO, p 110
74. Goldberg MC, Weiner ER (1989b) In: Averett RC, Leenheer JA, McKnight DM, Thorn KA (eds) Humic substances in the Suwannee River, Georgia: Interactions, properties, and proposed structures, U.S. Geological Survey, Open-File Report 87–557, Box 25425, Denver Federal Center, Denver, CO, p 179
75. Goldberg MC, Negomir PM (1989) In: Goldberg MC (ed) Luminescence applications in biological, chemical, environmental, and hydrological sciences, ACS Symposium Series No. 383, American Chemical Society, Washington, DC, chap 11
76. Goldberg MC (1982) The Sci of the Tot Env 24: 73
77. Imbrigiotta E Jr, Martin M, Sargent BP, Voronin LM (1989) In: Mallard GE, Ragone SE (eds) Geological Survey toxic substances hydrology program proceedings of the technical meeting. 26–30 Sept 1988 Phoenix Arizona pp 351–359. U.S. Geological Survey Investigative report 88–4220

Part 4. Fluorescence Immunoassay

17. Fluorescence Polarisation Immunoassay

Christian Klein, Hans-Georg Batz, Brigitte Draeger, Hans-Joachim Guder, Rupert Herrmann, Hans-Peter Josel, Ulrich Nägele, Roland Schenk and Bernd Vogt

Boehringer Mannheim GmbH, Werk Tutzing, Bahnhofstrasse 9-15, D-8132 Tutzing, Germany

1 Introduction

Since the introduction of radio immunoassay techniques [1], many different assay principles based on antibodies as specifiers have become routine tools in clinical diagnostics. Heterogeneous immunoassays, e.g. enzyme linked immunosorbant assays (ELISA) [2], require a series of time consuming incubation and wash steps, but they are characterized by a high sensitivity with a detection limit of 10^{-13} mol l^{-1} and even lower.

In contrast, homogeneous immunoassays are carried out without separation steps. The analyte and a labeled analyte (tracer) compete for binding to a specific antibody. A detectable physical or chemical property of the tracer is modulated by the immunoreaction. The modulated property may be e.g. enzyme activity, intensity of fluorescence, or fluorescence polarisation.

Examples for homogeneous immunoassay techniques based on photometric detection are enzyme-multiplied immunoassay technique [3] and cloned enzyme donor immunoassay [4]. Examples based on fluorescence detection are substrate-labeled fluorescence immunoassay (SLFIA) [5] and fluorescence polarisation immunoassay (FPIA).

The sensitivity of these assays is generally lower than heterogeneous immunoassays. One reason is that the competition principle precludes the use of a high molar excess of one of the immunoreactive reagents, like a solid phase bound antibody in ELISA.

2 Fluorescence Polarisation

The basic theory of fluorescence polarisation was developed by Perrin [6] in 1926. Weber [7] used this technique as a tool in structural investigations of the binding of small molecules to proteins. Dandliker and coworkers [8] established this method for the determination of thermodynamic and kinetic constants of the antigen–antibody reaction.

Later, Dandliker and colleagues [9] were the first to describe a fluorescence polarisation immunoassay (FPIA). When fluorescent molecules in solution are excited by polarized light, they emit partially polarized fluorescence. Polarization P is expressed by Eq. (1):

$$P = \frac{I_{\|} - I_{\perp}}{I_{\|} + I_{\perp}}.$$ (1)

$I_{\|}$ is the intensity of fluorescence passing a polarization filter parallel to incident electric field, $I_{\perp}$ is the intensity of fluorescence passing the polarisation filter perpendicular to the incident electric field (Fig. 17.1).

Following the theory of Perrin, Weber, Steiner [2, 3, 10], the polarisation P is a function of the rotational relaxation Q and the lifetime of excited state of fluorescence τ:

$$\left(\frac{1}{P} - \frac{1}{3}\right) = \left(\frac{1}{P_0} - \frac{1}{3}\right)\left(1 + \frac{3\tau}{Q}\right).$$ (2)

The rotational relaxation Q depends on the molecular volume V of the fluorescent molecule and the viscosity coefficient of the solution η:

$$Q = \frac{3\eta V}{RT}.$$ (3)

So polarisation is inversely related to the amount of Brownian motion. Polarisation increases with molecular volume V and decreases with lifetime of fluorescence τ.

In a fluorescence polarisation immunoassay a hapten-fluorochrome conjugate (tracer) and the free hapten from the sample compete for antibody binding sites. A free tracer with the molecular weight of normally lower than 1,000 Dalton (low molecular volume) rotates quickly and emits light with very low polarisation. The tracer–antibody complex has a high molecular weight of more than 150,000 Dalton: it rotates slowly resulting in a high degree of polarisation of the emitted fluorescence (Fig. 17.2).

With increasing amount of analyte present the polarization decreases, resulting in a declining calibration curve (see Figs. 17.9, 17.13)

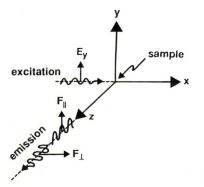

Fig. 17.1. Measurement of fluorescence polarisation

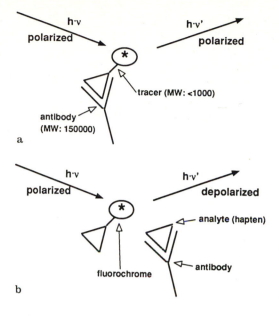

Fig. 17.2a, b. FPIA assay principle **a** without analyte: antibody bound to tracer rotates slowly, emitted light has maximum polarisation; **b** analyte present: analyte from the sample competes with tracer for antibody binding sites, free tracer rotates quickly, depolarisation of emitted light

3 Fluorophores

For maximum sensitivity a fluorophore used in FPIA should have a high extinction coefficient ε, high quantum yield Q_F and a high Stokes shift. A lifetime of fluorescence τ in the range of 4 ns results in an optimal polarization change from free to antibody-bound tracer.

A very popular fluorophore for use in FPIA has been fluorescein [11]. When excited at the fluorescein absorption maximum (485 nm), human sera exhibit a strong fluorescence, caused mainly by bilirubin and its oxidation products, and hemoglobin. Excitation above 550 nm drastically reduces serum fluorescence [12]. So red-shifted fluorophores absorbing >550 nm are promising candidates for FPIA.

Resorufin, a fluorophore absorbing at 572 nm, has been used in biochemistry e.g. for fluorogenic enzyme substrates [13]. To test the reduction of serum interferences by the use of resorufin we compared the serum background fluorescence of a serum pool with a high level of bilirubin (4.2 mg/dl) diluted 1:25. Excitation at 572 nm resulted in a fluorescence signal detected at 615 nm corresponding to 1.4 nmol l^{-1} resorufin tracer, while excitation at 485 nm gave a fluorescence at 525 nm equivalent to 7 nmol l^{-1} fluorescein tracer.

To detect low concentration analytes such as digoxin (see below) it is necessary to use low dilution of sample and low concentration of tracer for effective competition of the analyte. On the basis of the reduced serum background fluorescence a higher sensitivity was predicted for a resorufin-FPIA.

While rhodamines also allow excitation >550 nm they tend to bind nonspecifically to proteins. This nonspecific binding is critical because binding of a tracer through its fluorophore also results in high polarization values as does binding of a tracer through its hapten portion to an antibody.

We compared fluorescein, sulforhodamine 101, resorufin and substituted resorufins for binding to the critical protein human serum albumin (Fig. 17.3). While sulforhodamine 101 shows a strong polarisation when titrated with the albumin, fluorescein has the lowest affinity. Resorufin itself has a somewhat higher tendency to bind to serum albumin than fluorescein. But this property can be reduced by hydrophilic or ionic substitution to the level of fluorescein or even lower (Fig. 17.4).

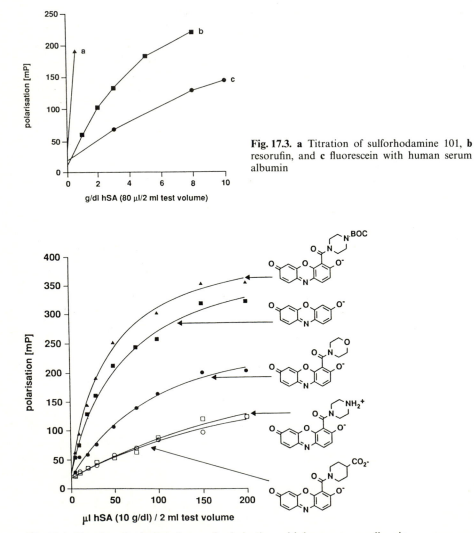

Fig. 17.3. a Titration of sulforhodamine 101, **b** resorufin, and **c** fluorescein with human serum albumin

Fig. 17.4. Titration of substituted resorufin derivatives with human serum albumin

Another basis for the use of a fluorophore in FPIA is the availability of activated compounds for coupling to hapten derivatives. The most popular chemistry is use of an amide bond derived from carboxylic acids and aliphatic amines. Examples of activated derivatives of fluorescein and resorufin are shown in Fig. 17.5. For coupling of fluorophores to amino group containing compounds as e.g. haptens or proteins we strongly recommend the succinimidyl ester derivatives like FLUOS or RESOS. Compared with isothiocyanates like

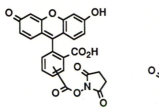
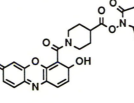

a   **FLUOS**   **RESOS**

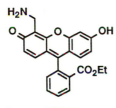
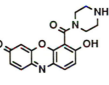

b   **AMFET**   **RESPIP**

Fig. 17.5. Activated derivatives of fluorescein and resorufin, **a** for the coupling to amines, **b** for the coupling to carboxylic acids. FLUOS: 5/6-carboxyfluorescein succinimidylester [14], RESOS: N-(resorufin-4-carbonyl)piperidine-4-carboxylic acid succinimidylester [14], AMFET: 2'-aminomethylfluoresceinethylester [15], RESPIP: resorufin-4-carboxylic acid piperazide [16]

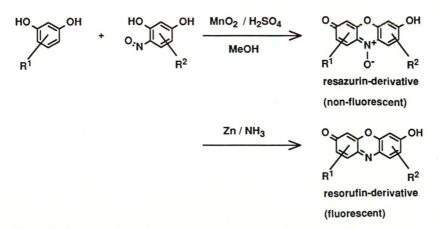

Fig. 17.6. General synthetic route to unsymmetrically substituted resorufins

FITC the succinimidylesters are characterised by a higher reactivity towards amino groups. FLUOS and RESOS allow shorter reaction times and a lower excess of reagent to yield high coupling rates. Reaction is possible at neutral or even slightly acidic pH [14].

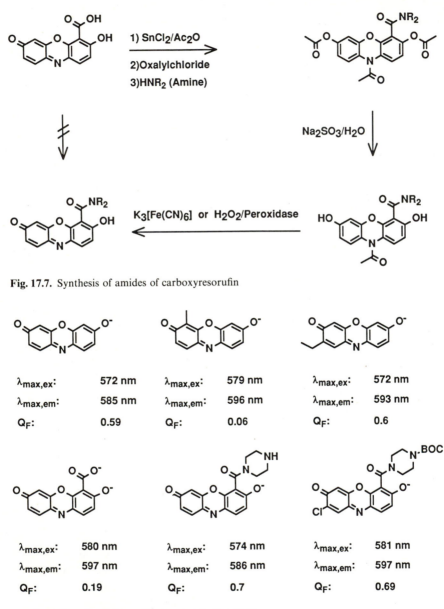

Fig. 17.7. Synthesis of amides of carboxyresorufin

10^{-6} mol l^{-1} solution in 0.1 mol l^{-1} phosphate buffer pH 7.5

Fig. 17.8. Spectral properties of substituted resorufins, 10^{-6} mol l^{-1} solution in 0.1 mol l^{-1} potassium phosphate buffer pH 7.5

Table 17.1. Characteristics of fluorescein, sulforhodamine 101 and resorufin relevant for FPIA, 10^{-6} mol l^{-1} solution in 0.1 mol l^{-1} potassium phosphate buffer pH 7.5

	fluorescein	sulforhodamine 101	resorufin
ε	75000	80000	64000
Q_F^a	0.87	0.86	0.59
$\lambda_{max, ex}$	492 nm	596 nm	572 nm
Stokes shift	26 nm	21 nm	13 nm
τ	4.6 ns	5 ns	3.9 ns
Activated derivatives	available	available	available
Unspecific binding	low	high	medium

^{a} 5×10^{-6} mol l^{-1} in 0.1 mol l^{-1} potassium phosphate buffer pH = 7.5

A variety of unsymmetrically substituted resorufins are easily synthesized under mild conditions (Fig. 17.6) allowing the incorporation of carboxy groups and other substituents controlling either excitation/emission wavelength or hydrophilicity [16].

While carboxy groups bound directly to the chromophore lead to strong quenching of fluorescence, the amides like RESOS or RESPIP have the same or even a better quantum yield Q_F than resorufin itself (Fig. 17.8). As shown in Fig. 17.7 the direct activation of the aromatic carboxylic acid and reaction with the suitable amine is not successful. The other way, via N,O,O-triacetyldi-hydroresorufins, yields the desired product. The key step here is the oxidative cleavage of the N-acetyl protective group with e.g. $K_3[Fe(CN)_6]$ or $H_2O_2/$peroxidase. Another critical class of substituents affecting the quantum yield are alkyl groups, but only if bound to position 4 (Fig. 17.8).

The comparison in Table 17.1 shows that resorufin is a promising alternative as fluorophore for the synthesis of tracers useful for FPIA [17].

4 Design and Screening of Tracers

Key criteria for the screening of the best antibody tracer combinations are
- standard-0-polarization as a measure of antibody affinity and ability to prevent depolarization in absence of the analyte
- dynamic range as a measure of effective competition of tracer and analyte for antibody binding
- dilution of the antiserum as a measure of affinity and/or content of antiserum "fitting" to the tracer.

Standard-0-polarisation is the maximum value of polarisation measured in the absence of analyte. The dynamic range is the difference between the standard-0-polarisation and the polarisation corresponding to the highest standard. An example is shown in Fig. 17.9, the calibration curve of a theophylline-FPIA. The concentration range to be determined of 0–40 $\mu g/ml^{-1}$ corresponding to 0–2.2 $\times 10^{-4}$ $mol\, l^{-1}$ is relatively high. So a high dilution of sample is possible for avoiding interference from serum compounds.

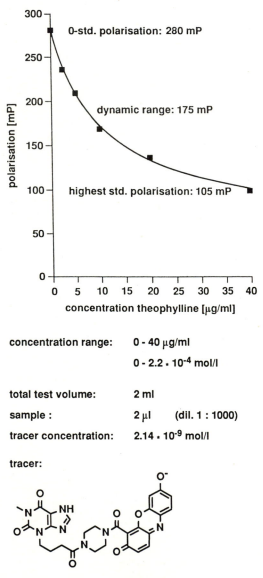

concentration range: **0 - 40 $\mu g/ml$**

                       0 - 2.2 · 10^{-4} mol/l

total test volume: **2 ml**

sample :   **2 μl (dil. 1 : 1000)**

tracer concentration: **2.14 · 10^{-9} mol/l**

tracer:

antibody: **pab<theophylline-3-immunogen>**

Fig. 17.9. Calibration curve of a theophylline-FPIA

Antibodies derived from theophylline-immunogens coupled through positions 3 or 8 have the desired ability to discriminate from similar compounds like caffeine. We tested the combinations of an anti-theophylline monoclonal antibody derived from a theophylline-8-immunogen and three sheep sera derived tbom a theophylline-3-immunogen with a variety of tracers for the best combination. The tracers (Fig. 17.10) differed in the coupling position of the fluorophore AMFET. Table 17.2 shows that the tracer coupled through position 3 fits best to one of the sheep sera derived from the theophylline-3-immunogen. The theophylline-8-tracer fits best to the mong lonal antibody derived from an immunogen of the same coupling chemistry. Both combinations have a high dynamic range, a high standard-0-polarisation and require low concentrations of the antibody. Nevertheless it is necessary to screen for the best serum obtained from an immunogen as examplified by the three different sheep sera.

Digoxin has the lowest concentration of the drugs usually monitored in serum with a concentration range of $0-6.4 \times 10^{-9} \ \text{mol}\,1^{-1}$.

In the test mix the concentration of the analyte and the tracer have to be in the same range to allow effective competition and a high dynamic range. The minimal tracer concentration is limited by serum background fluorescence. In the case of digoxin a low dilution of serum is necessary to allow a reasonably high tracer concentration.

We screened a series of tracers differing in length and structure of the spacer between the digoxigenin part and the fluorophore (Figs. 17.11, 17.12) for the

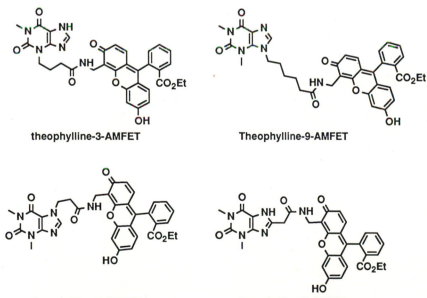

theophylline-3-AMFET   Theophylline-9-AMFET

theophylline-7-AMFET   theophylline-8-AMFET

Fig. 17.10. Structure of theophylline-tracers

Table 17.2. Screening for optimal anti-theophylline antibody/theophylline-tracer combination

Tracer	Monoclonal antibody ⟨theo-8⟩			Polyclonal sera ⟨theo-3⟩ Sheep A		Sheep B			Sheep C			
	Dyn. range	0-std.	Dil.	Dyn. range	0-std.	Dil.	Dyn. range	0-std.	Dil.	Dyn. range	0-std.	Dil.
theo-3-AMFET	—	—	—	120 mP	242 mP	1:10	230 mP	319 mP	1:40	240 mP	350 mP	1:46
theo-7-AMFET	—	—	—	230 mP	283 mP	1:1	19 mP	130 mP	1:1	—	—	—
theo-9-AMFET	—	—	—	199 mP	215 mP	1:1	—	87 mP	1:1	—	85 mP	1:1
theo-8-AMFET	210 mP:	328 mP	1:100	93 mP	143 mP	1:1	—	100 mP	1:1	81 mP	170 mP	1:1

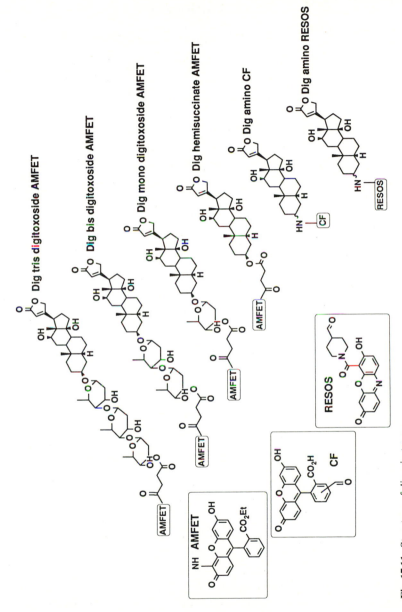

Fig. 17.11. Structure of digoxin-tracers

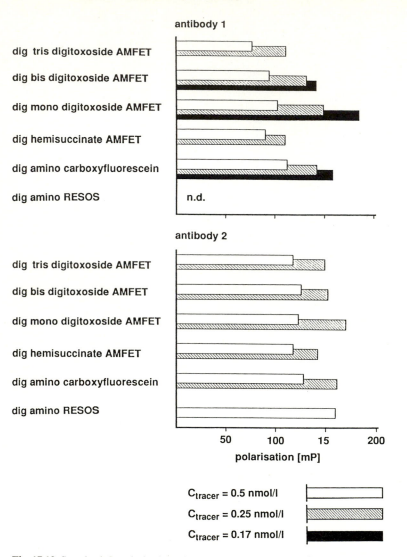

Fig. 17.12 Standard-O-polarisation of digoxin-tracers in combination with two antibodies

highest standard-0-polarisation. Generally, the lowest concentration of the tracer results in the highest standard-0-polarisation. Antibody 1 shows a pronounced difference in the ability to inhibit depolarisation of tracer fluorescence depending of the amount of digitoxose sugar units in the spacer. From other experiments we know that antibody 2 binds to the digoxigenin-part of digoxin, but not to the digitoxoses. So one could expect that the tracer with the longest spacer between the digoxigenin part and the fluorophore, "dig tris digitoxoside AMFET", has enough rotational freedom to emit depolarised fluorescence even when bound to the antibody.

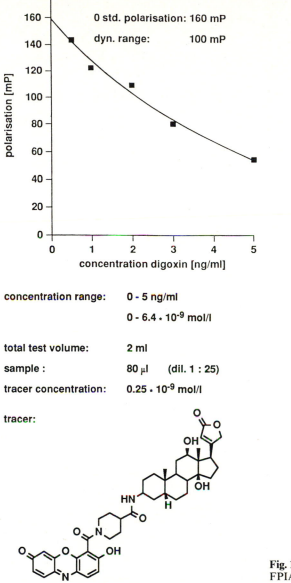

0 std. polarisation: 160 mP

dyn. range: 100 mP

y-axis: polarisation [mP]

x-axis: concentration digoxin [ng/ml]

concentration range: 0 - 5 ng/ml

0 - 6.4 $\cdot$ 10^{-9} mol/l

total test volume: 2 ml

sample :   80 µl   (dil. 1 : 25)

tracer concentration: 0.25 $\cdot$ 10^{-9} mol/l

tracer:

Fig. 17.13. Calibration curve of digoxin-FPIA

Nevertheless, Fig. 17.12 shows no drastic difference in standard-0-polarisation of this tracer as compared with others. Again it is necessary to screen for the best combination of tracer and antibody.

With the resorufin tracer "Dig amino RES" a digoxin-FPIA is made possible without sample pretreatment (Fig. 17.13). A fluorescein based digoxin-FPIA needs pretreatment steps to precipitate serum components [18].

Generally FPIA requires very small amounts of the tracer – e.g. 1 mg of the tracer shown in Fig. 17.13 is sufficient for more than 1,000,000 tests.

5 Summary

FPIA is a homogeneous assay principle based on simple reagents: unmodified antibody and chemically well defined and stable tracer. Costs of tracer and antibody are low because of their economical use.

Resorufin and fluorescein are both suited for use in FPIA on the basis of their spectroscopic characteristics and their low tendency to bind nonspecifically to proteins. Use of resorufin allows a higher sensitivity because of the lower serum background fluorescence.

The detection limit of FPIA in the range of 10^{-10} mol l^{-1} is sufficient for testing therapeutic drugs and drugs of abuse.

Disadvantages of FPIA as compared with e.g. cloned enzyme donor immunoassay are the declining and non-linear calibration curves. Furthermore FPIA affords equipment of the analyser with two polarisation and a fluorescent detection unit whereas the high through-put analysers currently used in clinical chemistry are based on simple photometric detection.

6 References

1. Yalow R (1978) Science 200: 1236
2. Engvall E, Jonsson K, Perman P (1971) Biochim Biophys Acta 251: 427
3. Rubenstein KE (1972) Biochem Biophys Res Commun 47: 846
4. Henderson DR, Friedman SB, Harris JD, et al. (1986) Clin Chem 32: 1637
5. Burd JF, Wong RC, Feeney JE, Carrico RJ, Boguslasky RC (1977) Clin Chem 23: 1402
6. Perrin F (1926) J Phys Radium 7: 390
7. Weber G (1953) Adv Protein Chem 8: 425
8. Dandliker WB, Feigen GA (1961) Biochem Biophys Res Commun 5: 299; Dandliker WB, Levison SA (1967) Immunochemistry 5: 171
9. Dandliker WB, Kelly RJ, Dandliker J, Farquah J, Levin J (1973) Immunochemistry 10: 219
10. Steiner RF, McAlister AJ (1957) J Polym Sci 24: 105
11. Jolley ME, Stroupe SD, Wang CJ, Panas HN, Keegan CL, Schmidt RL, Schwenzer KS (1981) Clin Chem 27: 1190; Jolley ME, Stroupe SD, Wang CJ, La-Steffes M, Hill HD, Popelka SR, Holen JT, Kelso DM (1981) Clin Chem 27: 1575
12. Dandliker WB, Hsu ML, Levin J, Rao BR (1981) Meth Enzymol 74: 3
13. Hofmann J, Sernetz M (1984) Analytica Chimica Acta 163: 67
14. Herrmann R, Josel H-P, Wörner W, Fetterhoff TJ (1989) 19th FEBS Meeting, Rome, July 2–7; Biochemica Dienst, Boehringer Mannheim No. 76, p 11 (1989)
15. Shipchandler MT, Fino JR, Klein LD, Kirkemo (1987) Anal Biochem 162: 89
16. Herrmann R, Klein C, Josel HP, Batz HG (1988) In: Yoshida Z, Kitao T (eds) Chemistry of functional dyes, Mita Press, Tokyo, p 87
17. USP 4.954,630 Boehringer Mannheim GmbH (1990)
18. Oeltgen PR, Shank WA, Blouin RA, Clark T (1984) Ther Drug Monit 6: 360

18. Progress in Delayed Fluorescence Immunoassay

Ilkka Hemmilä

Wallac Biochemical Laboratory, P.O. Box 10, SF-20101 Turku 10, Finland

Abbreviations

β-NTA	β-naphthoyltrifluoroacetone
PTA	pivaloyltrifluoroacetone
TMP-DPA	4(2′,4′,6′-trimethoxyphenyl)-2,6-dipicolinic acid
TOPO	tri-n-octylphosphine oxide
BCPDA	4,7-bis(chlorosulfophenyl)-1,10-phenanthroline-2,9-dicarboxylic acid
DTPA	diethylenetriaminepentaacetic acid
TR-FIA	time-resolved fluoroimmunoassay

1 Time-Resolved Fluorometry

The utilization of delayed detection of the emission in fluoroimmunoassays relates to the desire to improve the analytical sensitivity by means of temporal background rejection. Background is present in all fluorometric determinations and is due, among other things, to light scattering, emissions from the samples' endogenous fluorescence, autofluorescence of cells and tissues, luminescent properties of solid matrixes, cuvettes, test tubes, lenses, etc. Time-resolved fluorometry with pulsed excitation can be used to eliminate the interfering background, provided that the decay time of the specific signal clearly differs from that of the background (Fig. 18.1). The efficient use of time-resolved detection requires the employment of delay times longer than 10 µs, and, accordingly, luminescent probes exhibiting excited state life times longer than 10 µs must be developed for the system.

2 Long-Decay-Time Luminescent Probes

Various types of probes exhibiting photoluminescence of long emission duration have been tested in time-resolved fluoroimmunoassays [1], including probes emitting normal or delayed fluorescence, phosphorescent compounds and a variety of chelates exhibiting intra-chelate energy transfer sensitized luminescence (Table 18.1).

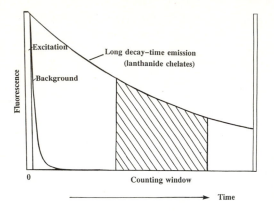

Fig. 18.1. The principle of pulsed excitation time-resolved fluorometry with delayed detection of long decay-time emission

Table 18.1. Probes used in time-resolved fluoroimmunoassays

Probe	Luminescence type	τ	Ref
Pyrene derivatives	Fluorescence	100 ns	[2]
Eosin	Delayed fluorescence		[3]
Erythrosin	Phosphorescence	250 μs	[4]
Metalloporphyrins	Phosphorescence	1–2 ms	[5]
Lanthanide chelates	Sensitized luminescence	1–2000 μs	[1]
Ruthenium chelates	CT sensitized luminescence	01.–7 μs	[6]

2.1 Chelate Labels

The only group of long life-time photoluminescent labels successfully applied in TR-FIA's so far are chelates, especially those of rare earth metals [1]. The luminescent properties of some lanthanide chelates are particularly well-suited for time-resolved fluoroimmunoassays requiring high sensitivity. In such chelates the wide absorptivity of the excitation light by the organic ligand at near UV ensures efficient light collection, whereas the chelated central ion collects the absorbed energy and produces a strong, narrow banded line-type emission at long wavelength, well-distinguished from the main part of the disturbing background. The exceptionally long luminescence decay time of these chelates allows the efficient use of the time-resolved detection principle.

3 Time-Resolved Fluoroimmunoassays

Based on the way the lanthanide chelates are used and according to the type of chelates applied, the published technologies can be divided into three major categories: fluorescence enhancement assays, chelate stabilization methods (or ligand saturation methods), and assays relying on stable fluorescent chelates. The two major technologies in routine use at the moment are DELFIA of

Wallac Oy (Turku Finland) [7] and FIAgen of CyberFluor (Toronto, Canada) [8]. The newly introduced chelate substrate of alkaline phosphatase can even expand the use of time-resolved fluorometry also to the field of enzyme detection, e.g. in EIA's [9].

3.1 Fluorescence Enhancement Based Assays

In assays applying fluorescence enhancement, the chelates used for labeling the immunoreagents are totally or partly different from those used in the fluoro-metric detection. In DELFIA, the antibodies (or in some competitive assays, antigens) are labeled with Eu^{3+} (or Sm^{3+}) by employing the non-fluorogenic bifunctional chelating agent, N^1-[p-isothiocyanatobenzyl]-diethylenetriamine-N^1,N^2,N^3,N^3-tetraacetic acid (Fig. 18.2). This is a small, hydrophilic, negatively charged molecule which gives the labeled reagent the stability needed during storage and use in immunoassays. After completion of the immunoreaction the chelated ion is dissociated, e.g. from the solid-phase bound antibodies, and converted into a new, highly fluorescent chelate with β-diketones (2-naphthoyl-trifluoroacetone) and TOPO, and measured in a micellar aqueous solution (Fig. 18.3). When Tb^{3+} or Dy^{3+} are used as the label ions, aliphatic β-diketones instead of aromatic ones have to be used in order to obtain an energy transfer from the ligand to the metal [7].

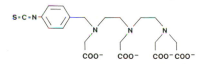

Fig. 18.2. Structure of the bifunctional chelating agent used for labeling antibodies with fluorogenic lanthanide ions

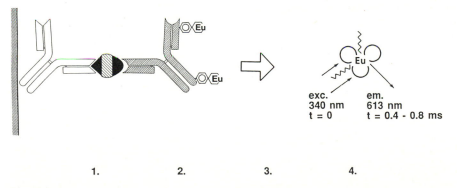

Fig. 18.3. Principle of DELFIA-type of assay with immunoreaction on solid surface (*1*) with Eu-labeled antibodies (*2*), use of dissociative fluorescence enhancement (*3*) and time-resolved fluorome-try (*4*)

The general advantages of the DELFIA technology include the high quantum yield (70%) of the formed fluorescent chelate in the micellar solution [1], the optimized optical arrangement (simple measurement from solution), the optimized hydrophilic labeling reagent, the applicability of the system to all possible multiple label approaches (biological or synthetic polymers) and dual-, triple- and quadruple-label assays. Regardless of the high quantum yield, the signal level and detection sensitivity can still be improved by taking advantage of an additional energy transfer process by utilizing a co-fluorescence effect [10]. The disadvantage of this technology, however, includes the potential restrictions caused by contamination risk, the extra step required for enhancement and the lack of spatial information (not suited for in situ assays nor for cytofluorometry).

3.2 Chelate Saturation Techniques

Saturation techniques are, in principle, the converse of the enhancement system described above; the reagents are labeled with a fluorogenic ligand, not with the metal ion. The fluorescence is enhanced after immunoreaction by including a high excess of the ions (e.g. Eu^{3+}) in order to saturate the surface bound ligands. In the commercialized technique, FIAgen, the primary antibodies are bio-tinylated and streptavidin is used in the indirect assay mode after labeling with the fluorogenic Eu-chelator, EuroFluor (BCPDA; 4,7-bis(chlorosulfophenyl)-1,10-phenanthroline-2,9-dicarboxylic acid (Fig. 18.4). The indirect assay mode is used to avoid the unspecific chelation of the ligand by the metals originating from samples. In the final incubation the chelator is saturated with an excess of Eu-ions, the surface is dried and measured with a laser fluorometer [8].

This technology has the general advantage of being less sensitive to contamination. The universal reagents and multilabel polymers developed can be applied in different types of assays and the assay can also give spatial information. In an indirect assay this requires, however, additional incubations with the tracer and Eu^{3+}. The measurement of surface fluorescence is optically more demanding and sensitive fluorometry requires dried surfaces. Recently a very large complex of chelate-labeled thyroglobulin has been introduced in order to improve the assay sensitivity [11].

3.3 Assays with Stable Fluorescent Chelate Labels

Chelates which can combine high fluorescence, good stability and hydrophilicity will be able to expand the application area of time-resolved fluorometry into

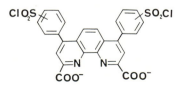

Fig. 18.4. Structure of the fluorogenic ligand, BCPDA (EuroFluor)

new fields. Different chelate structures have already been developed containing both suitable energy absorbing moieties (generally pyridine derivatives) and chelating structures (polyaminopolycarboxylates or cryptates) in the same compound [1]. These have been applied in homogeneous assays, e.g. for serum thyroxin [12], in heterogeneous solid phase assays, e.g. for hCG [13] and prolactin [14], in in situ quantitative immunofluorescence and hybridization imaging studies [15, 16] etc. When new fluorescent structures are discovered with improved solubility (hydrophilicity) and stability properties, these labels may be able to simplify the in vitro assay procedures as well.

4 Double-, Triple- and Quadruple-Label Assays

The narrow banded line-type emissions of the different lanthanides can also be exploited in measuring two or more chelate labels simultaneously from the same solution. As well as with emission filters, the signal distinction can be further improved by temporal resolution, because all the chelates exhibit their characteristic decay times in the time range from about 1 μs to about 1–2 ms (1) (Fig. 18.5).

The DELFIA technology can easily incorporate two different labels and after some modifications up to four labels have been measured from the same assay well. The composition of the enhancement solution can be stipulated for measurement of one, two, three or four labels simultaneously or sequentially. This opportunity has already been realized in double label immunoassays [17], and three- and four-label assays have also been demonstrated, e.g. by applying the co-fluorescence effect in the enhancement. Co-fluorescence enhancement is based on an inter-chelate energy transfer from the chelates of non-emitting ions

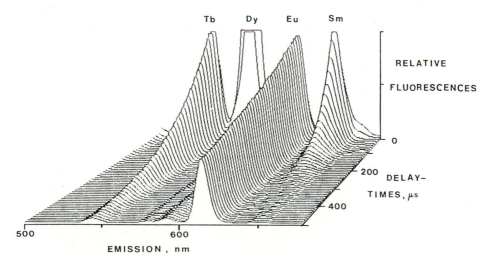

Fig. 18.5. The emission – decay profile of a mixture of Tb^{3+}, Dy^{3+}, Eu^{3+} and Sm^{3+} in PTA-based enhancement solution

Table 18.2. Fluorescence intensities (cps) and detection limits (pM) of Eu^{3+}, Sm^{3+}, Tb^{3+} and Dy^{3+} in various types of enhancement solutions using a time-resolved fluorometer

Ion	Ligand			
	β-NTA	PTA	TMP-DPA	PTA-Y^{3+}
Fluorescence of 1 nM Ln^{3+}:				
Eu^{3+}	1 000 000	252 800	ND	3 640 000
Sm^{3+}	13 000	1400	—	11 800
Tb^{3+}	—	140 000	675 000	1 880 000
Dy^{3+}	—	2000	—	19 500
Detection Limits:				
Eu^{3+}	0.05	0.42	ND	0.04
Sm^{3+}	3.5	24.0	ND	8.0
Tb^{3+}	ND	3.0	0.4	0.3
Dy^{3+}	ND	≈ 1000	ND	46.0

(e.g. Y^{3+}) to the chelates of emitting ions (Eu^{3+}, Sm^{3+}, Tb^{3+} and Dy^{3+}) producing greatly improved excitation efficiencies [10, 18].

Table 18.2 summarizes the relative fluorescence intensities and the respective detection limits obtained using various enhancement principles.

5 Homogeneous Assays

Homogeneous assays, that is where the analyte concentration is determined from the reaction mixture without any separations or washings, would particularly benefit from time-resolved fluorometry, because sample constituents (e.g. serum) impose a very high background problem in conventional fluorometry and fluoroimmunoassays. This actually sets the limits for the sensitivity obtainable. Homogeneous time-resolved FIA principles have also been demonstrated in numerous patents, but also in some clinical studies [12, 19]. It remains to be seen whether this technique will find more applications in areas such as therapeutic drug monitoring.

6 DNA Hybridization Assays

Assays based on DNA probes are complementary to immunoassays in certain specific areas, such as in the diagnosis of inherited diseases and detection of difficult microbes. Also in DNA hybridization assays the time-resolved fluorometry and the chelate labels form a technology strongly competing with the existing radioisotopic assays. Sensitivity levels earlier only possible with radioactive labels and autoradiographic detection have been achieved. The DELFIA-type technology has been applied in a number of hybridization assays, based on

Eu^{3+}-labeled streptavidin [20], directly labeled oligonucleotides [21] or poly-nucleotides [22]. Fluorescent and stable chelates have also been tested in DNA probe assays, e.g. by using macrocyclic Eu-cryptates [16], DTPA-salicylate chelate with Tb^{3+} [23] and bathophenanthroline chelated with Ru^{2+} [6].

7 Future Prospects

The performance of time-resolved immunoassays will be improved through improved labeling technologies, choice of better monoclonal antibodies, ap-plication of polymeric label clusters [11, 24], automation of the assay proced-ures, etc. The development of new chelate labels with improved properties will certainly expand the use of time-resolved technology also to totally new analytical fields. Progress will be expected in improved performances of immunoassays, expanding use of the technologies, e.g. in DNA probe assays, new applications in homogeneous assays and in cytofluorometric image ana-lysis.

One of the ideas inspired by the high sensitivity of time-resolved fluoro-metry and the spatial resolution of scanning laser microfluorometry is the multianalyte microspot assay design presented by Prof. Ekins [25]. On the other hand, the possibility of carrying out fluorometric analysis with two or three different time domains could be exploited in multianalyte imaging studies using the time domain between 1 and 100 ns for organic fluorescent probes, domain 100–1000 ns for medium-long decay time probes (e.g. pyrene-derivatives or the chelates of Dy^{3+} and Ru^{2+}) and time domain in the range from 50 to 2000 μs for the chelates of Eu^{3+}, Sm^{3+} and Tb^{3+}. The relatively slow energy transfer process in the chelates ensures that there is no temporal overlapping e.g. between the emissions of organic labels and the chelates [26].

8 References

1. Hemmilä I (1991) Applications of fluorescence in immunoassays. Wiley-Interscience, New York, NY
2. Wieder I (1978) In: Knapp W, Holubar K, Wick G (eds) Immunofluorescence and related staining techniques, Proc Sixth Int Conf Vienna, Elsevier, North-Holland Biomedical Press, 1978, p 67
3. Glick MR, Winefordner JD (1988) Anal Chem 60: 1982
4. Sidki AM, Smith DS, Landon J (1986) Clin Chem 32: 53
5. Savitsky AP, Papkovsky DP, Ponomaryot GV (1989) Dokl Acad Nauk SSSR 304: 1005
6. Bannwarth W, Schmidt D, Stallard RL, Hornung C, Knorr R, Müller F (1988) Helv Chim Acta 71: 2085
7. Hemmilä I (1988) Scand J Clin Lab Invest 48: 389
8. Diamandis EP (1988) Clin Biochem 21: 139
9. Evangelista RA, Pollak A, Gudgin Templeton EF (1991) Anal Biochem 197: 213
10. Xu Y-Y, Hemmilä IA (1992) Anal Chim Acta 256: 7
11. Christopoulos TK, Lianidou ES, Diamandis EP (1990) Clin Chem 36: 1497
12. Hemmilä I, Malminen O, Mikola H, Lövgren T (1988) Clin Chem 34: 2320
13. Hemmilä I, Båtsman A (1988) Clin Chem 34: 1163 (Abs. 054)

14. Mathis G, Amoravain M, Dedieu A, Socquet-Clerc F, Jolu EJP, Deschenaux R, Lehn J-M (1989) Abstract of Papers Presented at the IIIrd International Symposium on Quantitative Luminescence Spectrometry in Biomedical Sciences, Ghent, Belgium, May 1989.
15. Soini E, Pelliniemi LJ, Hemmilä IA, Mukkala V-M, Kankare JJ, Fröjdman K (1988) J Histochem Cytochem 36: 1449
16. Prat O, Lopez E, Mathis G (1991) Anal Biochem 195: 283
17. Hemmilä I, Holttinen S, Pettersson K, Lövgren T (1987) Clin Chem 33: 2281
18. Xy Y-Y, Pettersson K, Blomberg K, Hemmilä I, Lövgren T (1992) Clin Chem (in press)
19. Barnard G, Kohen F, Mikola H, Lövgren T (1989) Clin Chem 35: 555
20. Dahlén P (1987) Anal Biochem 164, 78
21. Dahlén PO, Iitiä AI, Skagius G, Frostell Å, Nunn MF, Kwiatkowski M (1991) J Clin Microbiol 29: 798
22. Hurskainen P, Dahlén P, Ylikoski J, Kwiatkowski M, Siitari H, Lövgren T (1991) Nucleic Acids Res 19: 1057
23. Oser A, Collasius M, Valet G (1990) Anal Biochem 191: 295
24. Lövgren T (1987) J Steroid Biochem 27: 47
25. Ekins RP, Chu F, Biggart E (1989) Anal Chim Acta 227: 73
26. Soini E, Hemmilä I, Dahlén P (1990) Ann Biol Clin 48: 567

19. Chemiluminescence Detection in Immunochemical Techniques. Applications to Environmental Monitoring

Eugène H.J.M. Jansen

Laboratory for Toxicology, National Institute of Public Health and Environmental Protection, P.O. Box 1, 3720 BA Bilthoven, The Netherlands

1 Introduction

Chemiluminescence is a phenomenon which has attracted increasing interest in recent literature not only because of its inherently great sensitivity, but also by the potentially widespread applications.

Chemiluminescence is the production of light by a (bio)chemical reaction. In contrast to fluorescence no external exciting light source is needed. This difference makes the chemiluminescence on some particular applications superior to fluorescence and results, for instance, in more simple detection systems. Since chemiluminescence is closely related to bioluminescence which is a natural phenomenon, a great number of applications are possible other than using chemiluminescence as a label system.

In the present report an overview is given of (potential) applications of chemiluminescence in immunochemical detection methods with possible applications in biochemical toxicology. Emphasis is put on the use of enzymes as label in immunoassays and in detection of specific proteins on western blots. In all cases the use of chemiluminescence will be considered critically and compared with other detection methods. In addition, a number of possible applications of chemiluminescence will be given in the field of environmental monitoring.

2 Chemiluminescent Labels in Immunoassays

The immunoassay is an analytical technique which is capable of specifically detecting very small quantities of a particular compound. In addition, a large number of samples can be detected simultaneously within a relatively short period of time (hours). The principles of an immunoassay will not be treated here. The most well-known immunoassay technique is the radioimmunoassay (RIA). For many years the RIA has proven to be a solid and versatile assay method for detecting small quantities of a given compound in complex biological matrices. However, the use of radioactivity is also one of the major drawbacks of the RIA. During the last ten years there has been a continuous search for alternatives, such as free radicals [1], fluorescent compounds [2], metals [3] and latex particles [4], rare earth metals [5], enzymes [6] and

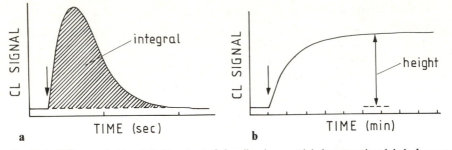

Fig. 19.1. Difference in kinetic light output of chemiluminescent labels. **a** transient labels. **b** enzyme labels

chemiluminescent compounds. The use of chemiluminescent compounds as labels in immunoassays has a number of additional advantages over radioactive labels. These advantages are: a higher sensitivity, an increased speed of detection and the advantages of having a non-isotopic immunoassay, such as no license needed for radioactive work, no radioactive waste disposal and the stability of the label.

The chemiluminescent labels can be divided into two classes mainly based on their kinetics of light output: the transient labels and the enzyme labels. As is shown in Fig. 19.1 the transient labels show a flash-like light output whereas the enzyme labels show a steady-state light output. In the next chapters the advantages and limitations of both kinds of label will be discussed in detail and the consequences with respect to sensitivity, practical use, and method of detection.

2.1 Transient Labels

2.1.1 Luminol and Derivatives

The best known chemiluminogenic compound is luminol [7] (Fig. 19.2a). Numerous examples of chemiluminescence immunoassays which use the iso-luminol system (in particular aminobutylethylisoluminol, ABEI) have been described (for reviews see [8–10]). Companies which have a commercial chemiluminescent immunoassay system based on luminol labels are Henning (Germany), Byk Mallinckrodt (Germany) and Bouty (Italy).

2.1.2 Acridinium Compounds

In the early 1980s, other very efficient chemiluminogenic labels were synthesized [11]. These so-called acridinium esters (Fig. 19.2b) can be oxidized by hydrogen peroxide to produce chemiluminescence.

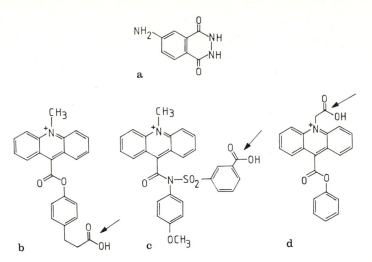

Fig. 19.2a–d. Chemical structures of chemiluminescent transient labels. **a** luminol; **b**, **c** and **d** acridinium compounds as explained in the text. The *arrows* indicate the site of attachment to haptens or antibodies

The main advantage of these kind of labels relative to luminol-like compounds is that there is no need for a catalyst. As a result very low background signals are observed with a consequently low limit of detection. Up until now only few applications have been reported of immunoassays using acridinium esters as label [12, 13]. The hydrophobicity of the acridinium moiety is probably the most important disturbing factor. Therefore, new kinds (second generation) of acridinium compounds have been developed and used in immunoassays [14, 15] (Fig. 19.2c). Recently, a third generation has been developed by Zomer et al. [16]. These *N*-functionalized acridinium compounds (Fig. 19.2d) can be coupled via the nitrogen of the acridine moiety to proteins which makes these labels suitable for homogeneous immunoassays by means of energy transfer. Companies which have a commercial chemiluminescent immunoassay system based on luminol labels are Ciba Corning (USA), Hoechst (Germany) and London Diagnostics (USA).

2.2 Enzyme Labels

The enzymatic reactions of horseradish peroxidase (HRP), xanthine oxidase (XOD) and alkaline phosphatase (AP) producing chemiluminescence are shown in Figs. 19.3a, b and c, respectively.

2.2.1 Horseradish Peroxidase

The enzyme labels consist of another class of labels which can produce chemiluminescence. In this case the catalyst is used as label instead of the

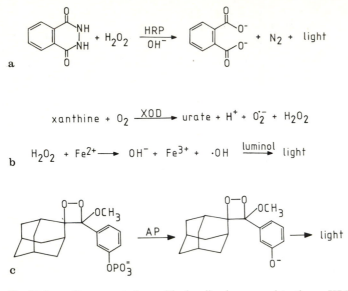

Fig. 19.3a–c. Enzyme reactions with chemiluminescence detection. **a** HRP; **b** XOD; **c** AP

chemiluminogenic compound. It turned out that the enzyme horseradish peroxidase (HRP) can be used very conveniently as a label in this kind of chemiluminescent reaction [17]. The introduction of phenolic enhancers in the signal reagent has a twofold positive influence on the system: the chemiluminescence signal produced by peroxidase is increased by a factor 100, whereas the blank signal is decreased [18]. A great number of very sensitive immunoassay have been reported using HRP as label [e.g. 19, 20]. Amersham (England) has developed the enzyme enhanced chemiluminescence into a successful commercial immunoassay system.

2.2.2 Xanthine Oxidase

Recently, another enzyme enhanced chemiluminescent system has been developed using the enzyme xanthine oxidase [21, 22]. This label system is superior to almost all existing chemiluminescent labels used in immunoassays with respect to sensitivity and kinetics. The system consists of the enzyme xanthine oxidase with hypoxanthine as substrate, an iron(II)-EDTA complex. This chelate splits hydrogen peroxide homolytically into hydroxyl radicals. These very reactive species react with luminol to yield chemiluminescence. The half-life time of the system at room temperature is about 30 h [22]. The long duration of the chemiluminescence signal is caused by the regeneration of the catalyst Fe(II)-EDTA in the Haber–Weiss reaction. In solution, the detection limit of xanthine oxidase is extremely low (3 attomol/tube). In the near future

Canberra Packard (USA) will probably be the first company to use xanthine oxidase in a commercial immunoassay system.

2.2.3 Alkaline Phosphatase

Other enzyme systems which are recommended for use as label in immuno-assays are enzymes which use not luminol but 1,2-dioxetanes as substrate, such as alkaline phosphatase and galactosidase [23, 24]. The substrates which are used are AMPPD and AMPGD, respectively, which can be activated to produce a prolonged chemiluminescent signal at 470 nm. Macromolecules, such as bovine serum albumin, amplify the chemiluminescence signal and improve the limit of detection by one order of magnitude. The very low detection limits (in the low attomol region) reported in the literature [24] could not be repeated in our laboratory. Two companies offer a signal reagent for chemiluminescence detection of alkaline phosphatase containing AMPPD (Tropix, USA; Lumigen, USA).

2.3 Advantages of Enzyme Labels

The advantages of enzyme labels described here have been summarized with respect to the chemiluminogenic transient labels. The main feature of the enzyme labels from which most of the advantages originate is the steady state kinetics: a long-term stable signal is produced for at least minutes or even hours.

The advantages of enzyme labels are the following. The reaction can be initiated outside the luminometer and after a certain time the samples can be placed inside the luminometer for counting. The measurements can be repeated if necessary. For xanthine oxidase and alkaline phosphatase the measurements can be repeated even the next day. Immunoassays in which enzymes are involved are very suitable for automation by using microtitre plate format devices. With microtitre plates (8 × 12 wells) 96 samples can be handled simultaneously and can be counted within one minute in a microtitre plate lumino-meter (Labsystems, Finland; Amersham, England). In fact, rather simple equipment can be used for the determination of the chemiluminescence. The luminometer must be able to detect a constant light signal. In contrast, the measurements of transient labels require a very efficient dispensing and mixing device but also a fast and advanced data collecting and processing system.

Because of the enzyme amplification and the enhanced chemiluminescence with the use of enzyme labels, a greater sensitivity and limit of detection can be achieved compared with transient labels. This can be illustrated with an example from practice if we consider two immunoassays of the anabolic steroid 19-nortestosterone (NT) with both an isoluminol-NT (NT-ABEI) and a HRP-NT label. To the assays 25 pg NT-ABEI and 100 pg NT-HRP were added as a label to obtain a sufficient high chemiluminescent signal. This corresponds to

15 pg and 0.75 pg NT equivalents on molar base, respectively. Thus, in the NT-ABEI assay, 20 times more label (based on NT-equivalents) is added than in the NT-HRP assay. Since the sensitivity of an immunoassay is proportional to the concentration oif the label, in principle the chemiluminescent enzyme immuno-assay (NT-HRP) will be about 20 times more sensitive than the chemilumin-escent immunoassay with luminol as transient label (NT-ABEI).

The commercial availability of mainly HRP- and AP-levels, already used in enzyme immunoassays with colorimetric detection, is rather good. In addition the in-house preparation of enzyme labels is rather simple.

2.4 Disadvantages of Enzyme Labels

The possible disadvantages of the use of enzyme labels with chemiluminescence detection in comparison with chemiluminogenic transient labels are the follow-ing. The stability of the enzyme preparations in solution can be a problem. Some enzymes, particularly HRP, lose their activity upon dilution [25]. Both the enzymatic reaction and the chemiluminescent enhancement reaction can be subject to interference caused by matrix components [22]. Therefore, it is recommended that enzyme immunoassays be performed in combination with solid phase techniques.

2.5 Comparison of Enzyme Labels

It is difficult to select one of the enzymes as a best label for immunoassays. In fact the three enzymes mentioned earlier (HRP, XOD and AP) can be used very conveniently as labels in enzyme immunoassays with chemiluminescence detec-tion. The characteristics of the enzymes which can be important for practical use in immunoassays with chemiluminescent detection are listed in Table 19.1. The non-linearity of HRP is in the region below 100 pg/tube. As a result, the limit of detection of HRP with colorimetric substrates (tetramethylbenzidine) is lower than with chemiluminescent substrates [26]. Very low detection limits were

Table 19.1. Characteristics of enzyme labels for use in chemilumines-cence immunoassays

Enzyme	Detection limit	Linearity	Kinetics	
			Lag time	Half-life
HRP	25 amol	− / +	1 min	1 h
XOD	3–18 amol	+	5 min	30 h
AP	3–13 amol	− / +	30 min	> 5 h

found for XOD [22] and AP [24]. The low detection limits for AP could not be repeated in our laboratory, however. Although there are no labels and immuno-assays commercially available, XOD seems to be the most promising enzyme label to use in combination with chemiluminescence detection.

3 Detection of Chemiluminescence

The light emitted by chemiluminogenic compounds or reactions can be meas-ured in a simple way and often with sufficient sensitivity using relatively simple equipment. In principle every detection device capable of measuring light can be used for this purpose.

The equipment to measure chemiluminescence is strongly dependent, how-ever, on the kind of reaction. In principle there are two kinds of label based on their kinetic behaviour. The transient labels, such as luminol and acridinium esters which show very fast kinetics in the order of seconds and the enzyme labels, such as horseradish peroxidase, xanthine oxidase and alkaline phospha-tase which show steady-state kinetics with a half-life in the order of minutes or even hours.

3.1 Luminometers

The measurement of the chemiluminescence of reactions with fast kinetics requires a number of extra features. First of all, the reagent which initiates the chemiluminescent reaction must be added in the measuring position before the photomultiplier. In addition, the injection system must be very efficient in order to achieve a complete mixing of the reagents. If the mixing is incomplete, an inhomogeneous solution is obtained in which the chemiluminescence reaction is also inhomogeneous and consequently not reproducible. The main manufac-turers of single-tube and automatic tube luminometers are Berthold, Lumac, Stratec, SLT, Turner, Bio-orbit, etc. [27].

Useful alternatives for field measurements are battery operated tube lumino-meters supplied by Dynatech, England (photocel) or Lumac, The Netherlands (photomultiplier). These devices cannot be used for the measurement of tran-sient labels because no facilities for the injection of reagents are present.

The essential components for the measurement and recording of the chemiluminescence of enzyme labels are a sample chamber and a light detecting device. The most suitable way of measuring the chemiluminescence of enzyme labels is in microtitre plate format. Recently, a chemiluminescence reader for microtitre plates has become available (from Labsystems, Finland) which is also capable of performing injections into the sample position [28] opening the possibility of measuring transient labels in microtitre plates.

3.2 Other Detection Devices

A good and cheap alternative for semi-quantitative chemiluminescence detection in microtitre plates is an X-ray film or a sensitive (20,000 ASA) instant photographic film [29, 30]. Although the photographic supports are less sensitive then luminometers, they can sometimes offer a good alternative.

Recently there has been some progress in the chemiluminescent detection of multiple samples by the use of a sensitive camera device [31, 32]. A successful approach is the use of image amplifiers to increase the sensitivity of CCD-cameras or the use of photon counting cameras. Coupling to an image processing system gives a very good sensitivity (equal to luminometers) which might be sufficient to detect chemiluminescent labels in immunoassays. Commercially available camera systems can be obtained from Hamamatsu Photonics (Germany) and Image Research (England).

4 Applications to Environmental Monitoring

Monitoring of environmental pollution can be performed by a variety of methods and with many kinds of sample. In this section a number of applications of environmental monitoring will be presented in which chemiluminescence plays an important role.

4.1 Immunoassays

As stated in the previous chapters immunoassays can be used very well to detect small quantities of a particular compound in large series. Chemiluminescence detection in these immunoassays can give additional advantages, like sensitivity and speed of detection. Another possibility of hapten-labelled (chemiluminescence) immunoassays is the manipulation of the specificity of the assay. On one hand a specificity can be achieved for only one compound and on the other hand a broad specificity can be developed for a whole group of related compounds or (unknown) metabolites of a particular compound. The limits of detection can go down to the low pg range/tube. In principle there is no upper range but concentrations in the high ng or low µg range/tube can be determined better by HPLC or GC. In any case the immunochemical detection can only be considered as a first screening technique. All positive findings must be confirmed by another independent analytical technique. Therefore emphasis must be put on avoiding false-negative results.

4.2 Biotransformation Enzymes

A well-known parameter for monitoring environmental contamination is the induction of cytochrome P-450 enzymes in the liver or other organs of animals.

The cytochrome P-450 system consists of a group of isoenzymes involved in the metabolism of both exogenic and endogenic compounds. The composition of the isoenzyme pattern of cytochrome P-450 in several organs is an important toxicological parameter, which can tell something about the kind of exposure to toxic compounds or environmental contaminants.

Besides enzymatic determinations which can only be performed on fresh liver samples, the presence and induction of cytochrome P-450 isoenzymes can also be detected and quantitated with Western blotting. Blotting is the transfer of macromolecules (proteins, DNA) from an electrophoretic gel to a membrane. By this procedure a copy of the gel pattern is obtained on the membrane. The protein pattern on the membrane can be used for analysis with specific antisera. Then the membrane is incubated with an antibody against a certain protein (primary antibody), followed by incubation with a so-called second antibody. This second antibody must be labelled with an enzyme in order to give a colour on the blot or to detect the chemiluminescence produced by the enzyme.

We recently compared the use and limits of detection of three enzymes (HRP, XOD and AP) coupled to the second antibody. The enzymatic activity on each blot was developed using colorimetric substrates or by chemiluminescence. As an example the detection of specific enzymes of the cytochrome P-450 biotransformation system in the liver of a rat treated with b-naphthoflavone has been used which has the same induction pattern as dioxin-like compounds. In Fig. 19.4 X-ray photographs of the chemiluminescence detection of HRP and

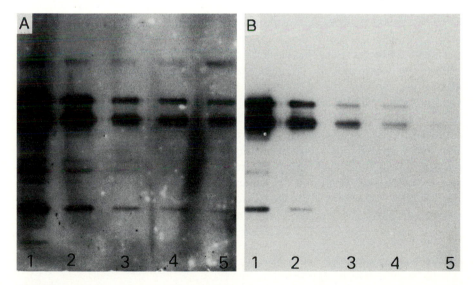

Fig. 19.4a, b. Western blots of microsomal protein in a three-fold diltion from left to right. Chemiluminescence detection was produced by enzymes coupled to a second antibody and registered on X-ray film **a** HRP with an exposure time from 10 to 30 min; **b** XOD with an exposure time from 1 to 24 h

Table 19.2. Experimental conditions and limits of detection of microsomal protein (expressed in ng protein) on Western blots (nitro cellulose) using enzymes couples to a second antibody

Enzyme	Detection	Substrate	Det. limit
HRP	Colorimetry	Diaminobenzidine	18
HRP	Chemiluminesc	ECL reagent	<18
AP	Colorimetry	BCIP	495
AP	Chemiluminesc	AMPPD	165
XOD	Chemiluminesc	Hypoxanthine + luminol	55

XOD are shown [33]. Remarkably, a high background staining is observed by chemiluminescent detection with HRP and AP, while on the chemiluminescent detection of XOD almost no background staining is observed.

From this study it appeared that detection of proteins on Western blots can be done in a sensitive way both by colorimetric and chemiluminescent detection. If very low limits of detection are required a chemiluminescent approach will be the method of choice. HRP showed the most sensitive chemiluminescence detection. XOD was also rather sensitive which can be improved further by longer exposure times. The main results of this study have been summarized in Table 19.2.

4.3 Luminous Bacteria

Another way of estimating water pollution is a simple test using luminous bacteria (photobacterium phosphoreum). When properly maintained and grown, the bacteria divert about ten percent of their metabolic energy into a special metabolic pathway that converts chemical energy into visible light by the luciferin–luciferase system. The presence of toxic substances that inhibits the cellular metabolism can be measured by a decreased light emission of the bacteria. Since the luminous bacteria are delivered in a freeze-dried form (Microtox, USA and Dr. de Lange, Germany), the toxicity tests can be performed with consistent sensitivity and reproducibility. A review of the relative sensitivity towards other biomonitoring tests for a number of organic and inorganic compounds has been published recently [34]. This test can also be used in mechanistic studies of the luciferin–luciferase bioluminescence system. In Fig. 19.5, an example is shown for the plasticizer diethylphthalate as measured in a microtitre plate Luminometer (Labsystems, Finland).

4.4 Hygiene Monitoring

Monitoring of hygiene, for instance in the food processing and food manufacturing industry, can be done by the detection of adenosine triphosphate (ATP) by the firefly luciferin–luciferase system [35]. When ATP is extracted from the

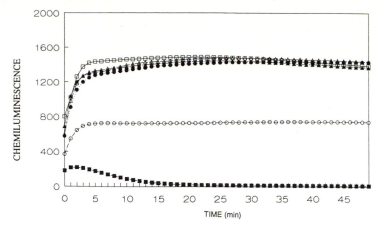

Fig. 19.5. Example of a toxicity test of diethylphthalate using luminous bacteria (Microtox). Chemiluminescence was measured in a microtitre plate (Labsystems). The added amounts were 0 (–□–), 12 (–▲–), 25 (–△–), 50 (–●–), 100 (–○–) and 200 μg (–■–) per well

sample and brought into contact with the firefly reagent, a reaction takes place with the production of light. In addition the origin of ATP, somatic or from microbial cells, can be established by different extraction methods. Besides a number of applications in biochemical toxicology such as the determination of ATP, ADP and AMP, a number of commercial applications can be obtained from Lumac, The Netherlands.

5 Conclusions

In conclusion, it will be clear that chemiluminescence, in general, is a very rapid and sensitive method of detection for a large number of applications. In the past emphasis has been put mainly on the use of chemiluminescent labels in immunoassays. From the foregoing it will be clear that the use of enzyme labels is the preference of the author. This preference has developed from practical experience with all the kinds of labels mentioned in the present report over a number of years. Sometimes, however, it is questioned if chemiluminescence will bring additional advantages over other detection methods such as the simple colorimetric immunoassays. Recently, there has been more commercial interest in the use of enzymes in combination of chemiluminescent detection on Northern and Western blots. Besides these applications a number of other applications originate from the fact that bio- or chemiluminescence is a natural phenomenon. Especially in biochemical toxicology, chemiluminescence as a technique is almost an obligatory tool for both research and routine measurements.

Acknowledgement. The author is indebted to many collegues who have contributed to chemiluminescence research for many years, especially to R. H. van den Berg and G. Zomer. Also the possibility, created by the Management of the National Institute of Public Health and Environmental Protection, of performing chemiluminescence research on various topics for several years is acknowledged.

6 References

1. Wei R, Almirez R (1975) Biochem Biophys Res Commun 62: 510
2. Sioni E, Hemmilae I (1979) Clin Chem 25: 353
3. Leuvering JWH, Thal PJHM, van der Waart M, Schuurs AHWM (1980) Fresenius Z Anal Chem 301: 132
4. Cambiaso CL, Leek AE, de Steenwinkel F, Billen J, Masson PL (1977) J Immunol Methods 18: 33
5. Lovgren TN (1987) J Steroid Biochem 27: 47
6. Schuurs AHWM, van Weemen BK (1977) Clin Chim Acta 81: 1
7. Albrecht HO (1928) Z Phys Chem 136: 321
8. Pazzagli M, Messeri G, Salerno R, Caldini AL, Tomassi A, Magini A, Serio M (1984) Talanta 31: 901
9. Wood WG (1984) J Clin Chem Clin Biochem 22: 905
10. De Boever J, Kohen F, Leyseele D, Vandekerckhove D (1990) In: Van Dyke K, Van Dyke R (eds) Luminescence immunoassay and molecular applications. CRC Press, Boca Raton, p 119
11. Weeks I, Beheshti F, McCapra F, Campbell AK, Woodhead JS (1983) Clin Chem 29: 1471
12. Weeks I, Sturgess ML, Siddle K, Jones MK, Woodhead JS (1984) Clin Endoc 20: 489
13. Richardson AP, Kim JB, Bernard GJ, Collins WP, McCapra F (1985) Clin Chem 31: 1664
14. Mayer A, Schmidt E, Kinkel T, Molz P, Neuenhofer S, Skrzipczyk HJ (1990) In: Stanley PE, Kricka LJ (eds) Bioluminescence and chemiluminescence. J Wiley, Chichester, p 99
15. Zomer G, Stavenuiter JFC, van den Berg RH, Jansen EHJM (1991) In: Baeyens WRG, De Keukeleire D, Korkidis K (eds) Luminescence techniques in chemical and biochemical analysis. Marcel Dekker, New York, p 505
16. Zomer G, Stavenuiter JFC (1987) Patent application O.A. 87.03075
17. Whitehead TP, Thorpe GHG, Carter TJN, Groucutt C, Kricka LJ (1983) Nature 305: 158
18. Thorpe GHG, Kricka LJ, Moseley SB, Whitehead TP (1985) Clin Chem 31: 1335
19. Thorpe GHG, Haggart R, Kricka LJ, Whitehead TP (1984) Biochem Biophys Res Commun 119: 481
20. van Look LJ, Jansen EHJM, van der Berg RH, Zomer G, Vanoosthuyze KE, van Peteghem CH (1991) J Chromatogr Biomed Appl 564: 451
21. Baret A, Fert V (1989) J Biolum Chemilum 4: 149
22. Jansen EHJM, van den Berg RH, Zomer G (1989) J Biolum Chemilum 4: 129
23. Clyne JM, Running JA, Stempien M, Stephens RS, Akhavan-Tafti H, Schaap P, Urdea MS (1989) J Biolum Chemilum 4: 357
24. Bronstein I, Edwards B, Voyta JC (1989) J Biolum Chemilum 4: 99
25. Jansen EHJM, van den Berg RH (to be published)
26. Jansen EHJM, van den Berg RH, Zomer G (1990) In: Van Dyke K, Van Dyke R (eds) Luminescence immunoassay and molecular applications. CRC Press, Boca Raton, p 57
27. Jago PH, Simpson WJ, Denyer SP, Evans AW, Griffiths MW, Hammond JRM, Ingram TP, Lacey RF, Macey NW, McCarthy BJ, Salusbury TT, Senior PS, Sidorowicz S, Smither R, Stanfield G, Stanley PE (1989) J Biolum Chemilum 3: 131
28. van den Berg RH, Jansen EHJM, Reinerink EJM, Zomer G (1990) In: Stanley PE, Kricka LJ (eds) Bioluminescence and chemiluminescence. J Wiley, Chichester, p 211
29. Thorpe GHG, Kricka LJ, Whitehead TP (1984) Clin Chem 30: 806
30. Jansen EHJM, Buskens CAF, van den Berg RH (1989) J Chromatogr Biomed Appl 489: 245
31. Jansen EHJM, Buskens CAF, van den Berg RH (1989) J Biolum Chemilum 3: 53
32. Hauber R, Miska W, Schleinkofer L, Geiger R (1989) J Biolum Chemilum 4: 367
33. Jansen EHJM, Laan CA, van den Berg RH, Reinerink EJM (1991) Anal Chim Acta in press
34. Munkittrick KR, Power EA, Sergy GA (1991) Environm Toxicol Water Quality 6: 35
35. Stanley PE (1989) J Biolum Chemilum 4: 375

Part 5. Fluorescence in Biomedical Sciences

20. Fluorescence Transients in Neurobiology: Applications of Voltage Sensitive and Ion Indicator Dyes

William N. Ross

Department of Physiology, New York Medical College, Valhalla, NY 10595, USA

1 Introduction

The use of time resolved fluorescence techniques in physiology has experienced a resurgence in recent years as a consequence of the synthesis of new indicator dyes responsive to interesting physiological parameters and the improvement of sophisticated computer-based imaging devices. The combination of these two developments has allowed physiologists to follow specific biological events in individual cells with high spatial and temporal resolution. This gives a global view of physiological events instead of the parochial perspective usually achieved with single site recording devices.

The opportunities offered by this technique extend into many areas of physiology, especially into studies of the nervous system where events occur on the millisecond time scale and with great spatial heterogeneity. In this system the difference between the view of a microelectrode recording the activity from one location within a single neuron and the view of an imaging system detecting events from many cells or many locations with a single cell is enormous and significant.

In this paper, I will give a short overview of applications using fluorescent indicators with particular reference to experiments that have been of interest to my laboratory. Three kinds of probes will be discussed: voltage sensitive dyes, sodium sensitive dyes, and in most detail, calcium sensitive dyes. The ion indicators change their excitation or emission spectra when they bind specific ions. Voltage sensitive dyes appear to work by changes in their spectra due either to changes in the viscosity or polarity of their microenvironment in the membrane or by shifts in the equilibrium between monomer and multimer forms of the molecules. With appropriate choice of filters these spectral changes can be detected as intensity changes and can be followed as a function of time. All of these indicators have been used to record events in different parts of individual neurons. In addition, voltage-sensitive dyes and calcium dyes have been used to monitor simultaneously the activity of large number of neurons.

There are a large number of voltage sensitive and ion indicator compounds. Many are fluorescent, but useful absorbance sensitive dyes also exist. The fluorescent dyes are preferred in applications where quantitative determinations of voltage or ion concentration are desired. "Ratio imaging" of calcium concentration levels is the best known example. Also the fluorescence technique is nicely matched to the current generation of low light level cameras. Experiments using absorbance sensitive dyes usually employ photodiodes or photodiode arrays because of the high light levels reaching the detectors. Their use is favored in applications where time resolution in the millisecond or faster range is needed. In experiments measuring transient changes of intensity they often give signals which can be detected with as high a S/N as when using fluorescent dyes.

2 Voltage Sensitive Dyes

In the early 1970s Cohen and his colleagues discovered that when neurons were stained with certain dyes the time dependence of the emitted fluorescence exactly followed the time dependence of the voltage changes of the stained cell [1]. They quickly realized that these measurements of light intensity changes could substitute for measurements of membrane voltage using a microelectrode in some experiments. Of particular interest were measurements from very small cells or parts of cells which were inaccessible to electrodes, or measurements from many cells where arrays of electrodes would be inconvenient.

One example of multicell imaging is the recording of action potentials from several neurons simultaneously in an invertebrate ganglion. In this kind of experiment the cells in the ganglion are stained by bathing the preparation in a solution containing the dye. After washing away the excess dye the fluorescence intensity from each cell is monitored as a function of time by elements of the photodiode array. The changes in fluorescence occur whenever the cells fire action potentials. After some computer processing the results can be displayed in a form which indicates the activity of each neuron. In some preparations up to 70% of the neurons can be followed simultaneously, suggesting that with sufficient analysis the behaviors that this ganglion controls might be understood at a level of completeness far beyond the capabilities of conventional techniques [2].

In another kind of experiment, a single cell is imaged with high magnification by the array in such a way that each photodiode element detects the voltage change from a small region of the neuron. When the fluorescence changes are recorded at high time resolution during neuron activity the propagation of voltages through the cell can be followed by noting the time delays of the fluorescence signals detected by each element (Fig. 20.1) [3]. This kind of information is useful in understanding how information is processed by individual neurons during synaptic integration. One limitation of these experiments is that current voltage sensitive dyes only give large signals when the cells are stained from the outside. Although there have been some efforts at developing dyes which can be injected into single cells [4] the sensitivity is not as great as obtained with topically applied dyes. Improvements in internal dyes would be very useful.

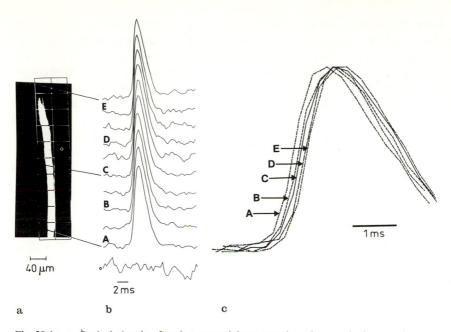

Fig. 20.1a–c. Optical signals of action potential propagation along a single axon in a nerve. **a** overlay of some of the photodiode elements used to detect the optical signals. The cell was stimulated with an electrode in the cell body. Lucifer yellow was injected into the axon and the nerve was stained with the voltage sensitive dye NK2367; **b** optical records from 10 elements along the axon. Note that the action potential is not detected by the photodiode element not located over the axon (–O–). 250 trials were averaged; **c** Five of these traces are shown on an expanded scale. The conduction velocity, calculated from the time at half amplitude, is 1.1 m/s. (From Ross, Krauthamer, 1984)

Low spatial resolution experiments have also been of interest e.g. where large parts of the nervous system containing thousands, or even millions of cells, are imaged onto the detector system [5]. Experiments approaching the nervous system in this way have recorded the spatial patterns on the visual cortex corresponding to light flashes into each eye and to oriented bars of light [6] to activation of individuals whiskers in rats [7], and to the delivery of specific odorants to the olfactory system [8]. Both photodiode arrays and cameras have been used profitably in these kinds of experiments.

3 Calcium Indicator Dyes

The use of ion sensitive dyes, particularly calcium sensitive dyes, in neurobiology and other areas of cell biology has been very fashionable in the last few years. The vast majority of studies measure the free intracellular calcium concentration following pharmacological or other biological interventions. This kind of experiment takes advantage of the fact that when the dye is excited at two wavelengths the ratio of the emitted fluorescence intensities can be calibrated in terms of the ion concentration [9]. In addition, a number of dyes have been synthesized which penetrate through cell membranes and, once inside, are transformed into the charged form by naturally occurring intracellular esterases. This property permits cells to be loaded with the indicator without injecting the dye through a microelectrode, a considerable advantage in many experiments. In contrast, membrane impermeant analogues can be directly injected into individual neurons embedded within a complex network. The optical measurements in these experiments reveal information only about that individual cell. The development of these dyes is largely the work of Roger Tsien and his colleagues, now at the University of California at San Diego.

We have used imaging of calcium changes in neurons to tell us information about several kinds of neurophysiological processes. The most common experiment has been to determine when and where a calcium dependent electrical event occurs in the neuron. Optical measurements at high time resolution show that there is usually a sharp jump in fluorescence corresponding to an increase in $[Ca]_i$. The location of these increases within the cell indicates where calcium came into the cell through the plasma membrane [10–12]. Making the measure-

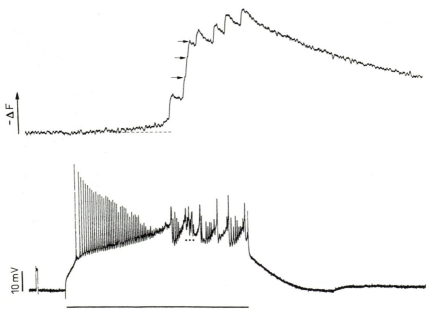

Fig. 20.2. Calcium transients detected from a small region of a Purkinje cell which was injected with the dye Fura-2. The electrical recording made from the soma shows an inactivating burst of fast Na spikes followed by a burst of complex spikes of mixed Na and Ca action potentials. The fluorescence transients, detected from the dendrites with a single photodiode with a response time of 2 ms, shows sharp decreases in fluorescence corresponding to each calcium action potential. Fluorescence excitation at 380 nm. (Modified from Lev-Ram et al., 1992)

ments at high time resolution allows us to distinguish membrane events, which occur rapidly, from the slower release of calcium from intracellular processes. Also, the faster the measurements are made the easier it is to separate influx from the redistribution of calcium which occurs later by diffusion [13].

One cell which we have studied extensively is the Purkinje neuron in the cerebellum. This cell, one of the largest in the CNS, has an extensive dendritic arborization which lies almost entirely in one plane, making it very suitable for optical experiments. When this cell is injected with a calcium indicator dye like Fura-2 sharp jumps in emitted fluorescence can be detected corresponding to individual calcium action potentials (Fig. 20.2). The jumps in fluorescence are detected mostly in the dendrites, with much slower, and usually smaller changes detected from the cell body (Fig. 20.3). We interpret this pattern to indicate that the calcium action potentials are generated in the dendrites. The slow changes in the soma correspond to bursts of fast action potentials [14, 15]. These conclusions match those suggested by previous electrophysiological recordings. One advantage of this new technique is that we can observe directly the spatial extent of the active firing process. In some cells we detected dendritic spiking whose

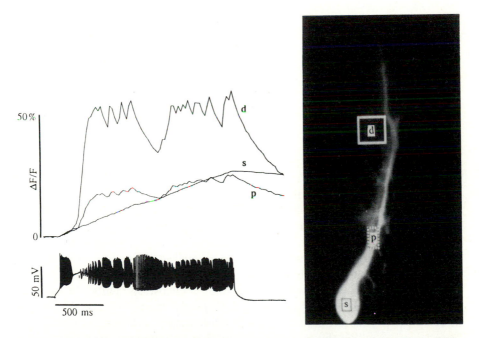

Fig. 20.3. Regional variation in the time course and amplitude of calcium transients in a cerebellar Purkinje cell. The image shows the cell filled with Fura-2. The rectangles marked with s, p, and d indicate the areas from which the time dependent traces are taken. The somatic signal (s) shows a slow buildup of $[Ca]_i$ due to the ongoing burst of fast sodium spikes. The distal dendritic signal (d) shows transients corresponding to the calcium spikes. The proximal region (p) has signal influenced by both kinds of action potentials. (Modified from Lev-Ram et al., 1992)

locus changed from spike to spike [16]. These kinds of patterns have important implications for the analysis of synaptic integration in Purkinje cells.

We have applied this technique to several different kinds of neurons in the CNS and in invertebrate preparations. In each case the calcium imaging experiments have revealed patterns of excitation which were not detected with electrode recordings.

4 Sodium Indicator Dyes

Tsien and his coworkers have also synthesized fluorescent dyes sensitive to changes in intracellular [Na] [17]. In the last year we have begun to exploit one of these dyes, SBFI, in a manner similar to that used for calcium sensitive dyes. In most neurons sodium ions are the major carriers of inward current during action potentials and synaptic potentials. Detection of $[Na]_i$ transients in localized regions of a neuron can be used to indicate where Na channels are distributed in the cell and where Na-dependent potentials propagate. For example, in the Purkinje cell the pattern of sodium entry reveals that Na channels are concentrated in the axon hillock, a region of the axon close to the cell body, at lower concentration in the soma, and almost absent in the dendrites (Fig. 20.4) [18]. This distribution contrasts with the dendritically dominated distribution of calcium channels determined with the calcium sensitive dye Fura-2. When both calcium and sodium sensitive dyes are used together new information can be produced. In the somatic region the distribution of transients evoked by fast sodium action potentials is different when detected with SBFI and with Fura-2. This shows that most of the calcium entry must be coming through different channels than sodium entry.

5 Ion Diffusion

When $[Ca]_i$ or $[Na]_i$ rise transiently large concentration gradients are usually produced. These result both from the fact that entry is at the surface of the cell causing an initial peak concentration just under the membrane, and also because entry is often localized. These initial concentration peaks subsequently are reduced by buffering, sequestration into organelles, and removal through the membrane. In addition, diffusion down the concentration gradient plays an important role in reestablishing intracellular homeostasis. Diffusion into the center of a cell has been demonstrated elegantly using high speed confocal imaging [19]. This technique detects time dependent changes in fluorescence from locations within the center of a cell without interference from light above and below the point of interest. Longitudinal diffusion away from localized sites of entry has been observed at synapses. In the barnacle photoreceptor, a favorable preparation for analyzing graded synaptic transmission, we have shown that diffusion is the main process removing calcium from the presynaptic

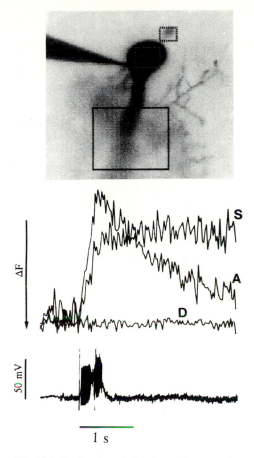

Fig. 20.4. Regional variation in sodium transients corresponding to an intrasomatically elicited burst of fast Na spikes. The largest signal is detected from the axon hillock (*A*). It recovers rapidly due to Na diffusion into the soma. The somatic signal shows no sign of recovery during the time of this trace. The absence of signal over the dendrites is consistent with a low density of Na channels in this region and the consequent failure of the Na spikes to propagate out of the soma. (Modified from Lasser-Ross and Ross, 1992)

terminal region and consequently a standing $[Ca]_i$ gradient is established in the cell when it is depolarized [20].

Calcium diffuses relatively slowly within neuronal cytoplasm because of the high concentration of intracellular buffers and sequestering organelles. Sodium diffusion is much more rapid, with a diffusion rate approaching that of the ion in water. Consequently, diffusion plays an even more significant role in establishing equilibrium after a transient increase in $[Na]_i$. In Purkinje cells this process is responsible for the rapid recovery in the axon hillock following stimulation.

6 Imaging Technology

Two kinds of imaging devices have been used in optical recording experiments. The first is an array of photodiodes arranged in a square (usually 10×10 or 12×12 with a 24×24 array just becoming available). Arrays usually have individual amplifiers connected to each element which permits sensitive measurements of small changes in light intensity and high speed sampling of each element (1 kHz or higher is typical) albeit with low spatial resolution. Imaging systems based on these arrays have been used most profitably for recording the very fast signals associated with voltage sensitive dyes [21] and absorbance sensitive calcium dyes [11].

The second kind of device is a camera, either a high sensitivity, tube-based SIT camera or a scientific grade cooled CCD camera (e.g. Photometrics, USA). The CCD camera can be programmed to take images at 100 frames/s at low spatial resolution while preserving high dynamic range, great sensitivity, accuracy, and millisecond response time [22, 12]. We have used this kind of camera to record the dynamics of intracellular ion concentration changes in individual neurons. The 10 ms frame interval is adequate to track the time course of most physiological events.

An often neglected part of camera systems used to follow physiological events is appropriate and efficient computer software to integrate imaging and electrical measurements and to analyze the data to recover useful biological data. Image processing packages, often developed for industrial or anatomical applications, are not very useful.

7 Future Directions

From the point of view of a physiologist we can see these kinds of fluorescence measurements improving in four different ways:
(1) Better dyes – Although the current generation of calcium indicators dyes is satisfactory for many applications, improvements would help certain experiments. Dyes with higher quantum yield would improve measurements of rapid, small amplitude transients from very small processes. Dyes with a higher K_d are necessary for making measurements in environments where $[Ca]_i$ is high, like muscle cytoplasm or the extracellular space. Dyes with different excitation bands would be useful in cells where the near UV excitation spectrum of Fura-2 overlaps with tissue autofluorescence or the absorption spectrum of intrinsic pigments. Several dyes in this category have recently been synthesized and are available from Molecular Probes. Of particular interest would be analogues of current probes which were specifically sequestered or excluded from various intracellular compartments or possibly even from specific cell types. Improvements in the sensitivity of voltage sensitive dyes are clearly needed.

(2) Different kinds of indicators – The current repertoire of useful physiological probes includes sensors for Ca, Na, Mg, pH, cAMP, and membrane potential. Indicators for other intracellular and extracellular messenger molecules like IP_3 or hormones would also be useful.

(3) Instrumentation improvements – The current generation of silicon based detectors, both photodiodes and CCD cameras, have almost ideal quantum efficiency, linearity, and fast response time. Systems based on photodiode arrays have high time resolution and low spatial resolution, CCD cameras the reverse. Improvements in the speed and sophistication of software controlling the acquisition and analysis of physiological experiments are to be expected considering current directions in the computer industry. One trend to watch is the use of confocal microscopes to record time dependent fluorescence changes in three dimensions. Current devices do not yet have sufficient sensitivity and/or low enough noise to follow events within individual cells without signal averaging. But improvements in laser amplitude stability and positioning accuracy are likely to improve this situation.

(4) New kinds of experiments – This category is hardest to predict. But two directions are clear. One is the application of these techniques to smaller and more diverse kinds of cells or cell processes. Almost every cell which has been examined in this way has revealed interesting new physiology. This trend will continue as the technique becomes more refined. The second involves the detection of physiological signals from many cells simultaneously. If this can be accomplished with enough temporal resolution it will be possible to analyze networks of cells with a level of precision not previously possible.

Acknowledgements. Supported in part by USPHS grant NS16295 and NSF grant BNS8819188. Development of the CCD camera system was supported by the Whitaker Foundation. Experiments from the author's laboratory were done in collaboration with Joseph Callaway, Nechama Lasser-Ross, Varda Lev-Ram, Hiroyoshi Miyakawa, Ann Stuart, and Robert Werman.

8 References

1. Cohen LB, Salzberg BM, Davila HV, Ross WN, Landowne D (1974) J Memb Biol 19: 1
2. Cohen L, Hopp H -P, Wu J -Y, Xiao C, London J, Zecevic D (1989) Ann Rev Physiol 51: 527
3. Ross WN, Krauthamer V (1984) J Neurosci 6: 1148
4. Grinvald A, Salzberg BM, Lev-Ram V, Hildesheim R (1987) Biophys J 51: 643
5. Grinvald A, Frostig RD, Lieke E, Hildesheim R (1988) Physiol Rev 68: 1285
6. Blasdel GG, Salama G (1986) Nature 218: 579
7. Orbach HS, Cohen LB, Grinvald A (1985) J Neurosci 5: 1886
8. Kauer JS (1988) Nature 331: 166
9. Grynkiewicz G, Poenie M, Tsien RY (1985) J Biol Chem 260: 3440
10. Ross WN, Stockbridge LL, Stockbridge NL (1986) J Neurosci 6: 1148
11. Ross WN, Werman R (1987) J Physiol Lond 389: 319
12. Lasser-Ross N, Miyakawa H, Lev-Ram V, Young SR, Ross WN (1991) J Neurosci Methods 36: 253
13. Ross WN (1989) Ann Rev Physiol 51: 491
14. Lev-Ram V, Miyakawa H, Lasser-Ross N, Ross WN (1992) J Neurophysiol (in press).
15. Miyakawa H, Lev-Ram V, Lasser-Ross N, Ross WN (1992) J Neurophysiol (in press).

16. Ross WN, Lasser-Ross N, Werman R (1990) Proc Roy Soc B240: 173
17. Minta A, Tsien RY (1989) J Biol Chem 264: 19449
18. Lasser-Ross N, Ross WN (1992) Proc Roy Soc B 247: 35
19. Hernandez-Cruz A, Sala F, Adams PR (1990) Science 247: 858
20. Lasser-Ross N, Callaway J, Stuart AE, Ross WN (1991) Ann NY Acad Sci 635: 475
21. Cohen LB, Lesher S (1986). Soc Gen Physiol Ser 40: 71
22. Connor JA (1985). Proc Natl Acad Sci USA 83: 6179

21. Optical Monitoring of Postsynaptic Potential in the Early Embryonic Avian Brain Stem Using a Voltage-Sensitive Dye

K. Kamino, T. Sakai, Yoko Momose-Sato, H. Komuro, A. Hirota and K. Sato

Department of Physiology, Tokyo Medical and Dental University, School of Medicine, Bunkyo-ku, Tokyo 113, Japan

1 Introduction

The ontogenetic approach to generation of physiological events during natural development would be a useful and powerful strategy for studying central nervous systems: it would allow us to analyse progressively the complicated functional organization and architecture of nervous systems, in a manner reminiscent of the expansion of a complex function in a power series. However, the experimental analysis of early embryonic nervous systems is technically difficult because the cells are extremely inaccessible: the microelectrode examination of neural cells, which provides the most direct test of their electrophysiology, is often difficult because of the small size of the cells. For this reason, electrophysiological studies of very early developing embryonic nervous systems have been hampered.

In such a serious situation, optical methods for monitoring electrical activity using voltage-sensitive dyes have been introduced, and it has been possible to make recordings from embryonic neurons at very early stages of their development [1–6]. Here, some results obtained with optical recordings of synaptic activity in the early embryonic brain stem are briefly sketched.

2 Optical Methods for Monitoring Cellular Electrical Activity

Optical methods for monitoring cellular electrical events offer two principal advantages over conventional electrophysiological techniques. One is that it is possible to make optical recording from very small cells and subcellular organelles which are inaccessible to microelectrode impalement [7, 8], and another is that multiple regions of a preparation can be monitored simultaneously and provide spatially-resolved mapping of electrical activity [9–11]. Accordingly, early embryonic nervous systems are particularly suitable subjects for the optical method.

2.1 Voltage-Sensitive Dyes

In light of the fact that early embryonic heart and neural cells are primitive and may be very sensitive to small changes in the environment, we have tested several dyes. And, finally, we designed an improved merocyanine-rhodamine dye, NK2761 [11, 12, 13]. This dye was synthesized by Nippon Kankoh–Shikiso Kenkyusho Co., Okayama, Japan. It is an analogue of dye XVII [14] and XXIII [15, 16], in which the alkyl chain (ethyl) attached to the rhodanine nucleus is replaced by a butyl. This dye has several advantages in that (1) for staining the preparation, a lower concentration (0.05–0.2 mg/ml) is used; (2) neither the pharmacologic effects nor the photodynamic damage are serious; (3) dye bleaching time is relatively long; and (4) the optical response follows the voltage-change with a time constant of less than 2.0 μs. In addition, NK2761 is sufficiently soluble in physiological Ringer's solution.

2.2 Optical Recording System

The multiple-site optical recording system that we currently use consists of four main parts: optics, detection system, multiple-channel recording system and computer (Fig. 21.1). The preparation chamber is placed on the stage of an microscope mounted on a vibration-isolation table. Bright field illumination is provided by a tungsten-halogen lamp driven by a stable DC power supply. Incident light is collimated, passed through a heat filter, rendered quasi-monochromatic with an interference filter, and focused onto the preparation by means of an aplanatic/achromatic condenser. When merocyanine-rhodanine dyes are used, a 703 ± 15 or 702 ± 13 nm interference filter is employed.

An objective (S plan Apo) and a photographic eyepiece form a magnified real image of the preparation, at the image plane. The transmitted light intensity at the image plane of the objective and photographic eyepiece are detected using a 12 × 12 square matrix array of silicon photodiodes. Thus, spatial resolution is limited only by microscope optics and noise considerations. In the present experiments, using a × 10 objective together with a × 2.5 photographic eyepiece, the magnification of the image was usually 25 ×.

The outputs of the detectors in the diode array are fed to 144 amplifiers via 144 current-to-voltage converters. The amplified outputs from 128-elements of the diode array are first recorded simultaneously on a 128-channel data recording system, and then are fed into a computer. The programs are originally written in assembly language (Macro-11) called by Fortran under the RT-11 operating system (version 5.0).

3 Embryonic Chick Brain Stem Preparations

We used mainly embryonic chick brain stem slice preparations. Fertilized eggs of white Leghorn chickens were incubated for 4–9 days in a forced-draft incubator at temperature of 37 °C and 60% humidity, and were turned once each hour. The brain stems, with vagus nerve fibers attached, were dissected from the embryos. The pia mater attached to the brain stem was carefully

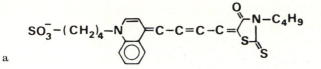

a

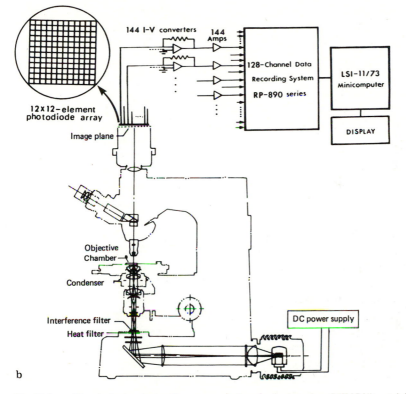

b

Fig. 21.1. a Chemical structure of a merocyanine-rhodanine dye (NK2761), and **b** schematic diagram of the multiple-site optical recording system that is currently used for monitoring changes in transmitted light from the stained preparation

removed in the bathing solution under a dissecting microscope (Fig. 21.2). Slices were then prepared, with the right and/or left vagus nerve fibers attached, by sectioning the embryonic brain stem transversely at the level of the root of the vagus nerve. The thickness of the slice was about 1 mm. The isolated brain stem preparation was attached to the silicone bottom of a simple chamber by pinning it with tungsten wires. The preparation was kept in a bathing solution with the following composition (in mM): NaCl, 138; KCl, 5.4; CaCl$_2$, 1.8; MgCl$_2$, 0.5; glucose, 10; and tris HCl buffer (pH 7.2), 10. The solution was equilibrated with oxygen.

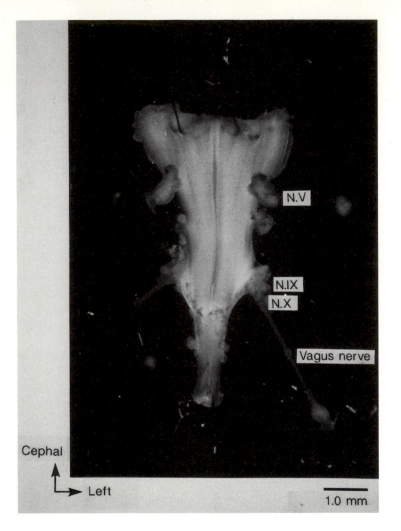

Fig. 21.2. Photomontage of the ventral view of a 7-day-old embryonic chick brain stem. This preparation was stained with the merocyanine-rhodanine dye, NK2761

The isolated preparation was stained by incubating it for 15–25 min in a Ringer's solution containing 0.1–0.2 mg/ml of the voltage-sensitive merocyanine-rhodanine dye NK2761, and the excess (unbound) dye was washed away with dye-free Ringer's solution before recording. For preparations in which the vagus nerve was stimulated, the cut end of the nerve was drawn into a suction electrode fabricated from haematocrit tubing, which had been hand-pulled to a fine tip (about 100 μm internal diameter) over a low-temperature flame.

4 Optical Detection of Evoked Action Potentials

Figure 21.3a shows an example of original recording of the optical signals obtained from an early 7-day-old embryonic intact brain stem preparation in response to vagus stimulation. The square current pulse of 40 μA and 5 ms in duration was applied on the right vagus nerve with a suction electrode. The preparation was about 700 μm in thickness and translucent. We therefore detected optical responses by means of transmitted light. The change in the light intensity were detected simultaneously from 127 adjacent regions of the stained brain stem preparation, using the 12×12-element photodiode array. This recording was made at 700 nm.

The fractional changes of these optical signals were 10^{-4} to 10^{-3}. The fractional change of the optical signal depends on the wavelength of the incident light. The action spectrum for the merocyanine-rhodanine dye (NK2761) is shown in Fig. 21.3b. The transmission intensity change was an increase between 500 and 620 nm, and a decrease between 640 and 750 nm. The largest signals were detected at 700–720 nm, and no signals were found at 630 nm.

The evoked optical signals were spike-like, and appeared to be concentrated in a limited area. These signals were completely blocked by tetrodotoxin of 20 μM. Therefore, it is likely that the evoked optical signals are due to Na^+ dependent action potentials [2].

5 Postsynaptic Potential Signals

Figure 21.4 illustrates two examples of optical recordings of neuronal activity in the embryonic brain stem slice from an 8-day-old embryo, in response to vagus stimulation. Positive square current pulse stimuli, which depolarize the axonal membrane, were applied to the vagus nerve fibers. The optical signals at 700 nm were detected simultaneously from 127 adjacent regions of the preparation using a 12×12-element photodiode array.

In Fig. 21.4, the optical signals evoked by a 4.2 μA/7.0 ms stimulus appeared on the dorsal surface of the brain stem, and apparently they consisted of two components: the spike-like fast signal and the succeeding long-duration slow signal. Enlargements of the signals detected from six different positions (H-3, I-3, L-4, K-7, H-10 and G-11) are shown on the bottom of this figure. The action spectra of these two components were the same and both components were completely eliminated at 620–630 nm, the null wavelength for the NK2761 voltage-dependent optical change. No optical changes were detected from unstained preparations. These results indicate that both the fast signal and the slow signal are indeed dye-absorption changes related to changes in membrane potential and do not correspond to changes in light scattering related to mechanical or other factors.

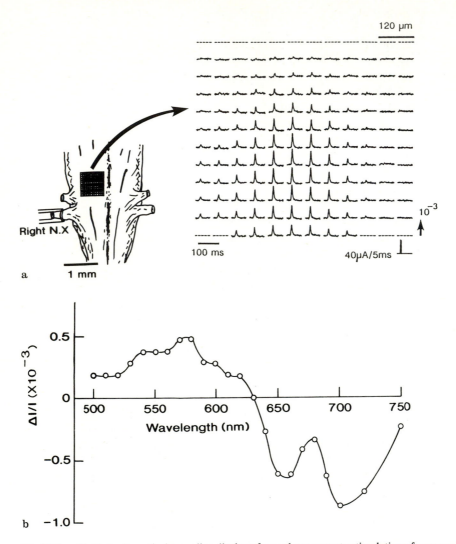

Fig. 21.3. a Multiple-site optical recording display of neural responses to stimulation of vagus nerve fibers in an early 7-day-old embryonic chick brain stem preparation. The preparation was stained with a merocyanine-rhodanine dye (NK2761). The signals were detected from the ventral side of the brain stem, with a 702 ± 13 nm interference filter using the 12 × 12-element photodiode array. The traces represent signals detected by one photodiode from a 56 × 56 μm² area of the brain stem. The traces are arranged so that their relative positions correspond to the relative positions of the area of the preparation imaged onto the detectors. The outputs of the individual detectors have been normalized by dividing by the DC background intensity. Eight trials were averaged. The direction of the *arrow* at the lower right corner of the optical recordings indicates a decrease in transmission (increase in absorption) and the length of the arrow represents the stated value of the fractional change. The recording was obtained with 40 μA/5 ms depolarizing current applied to the right vagus nerve fibers with a suction electrode. The relative location of the photodiode array on the image of the brain stem is illustrated on the left. **b** Wavelength dependence of the absorption changes resembling action potential, in the 7-day-old embryonic brain stem stained with NK2761

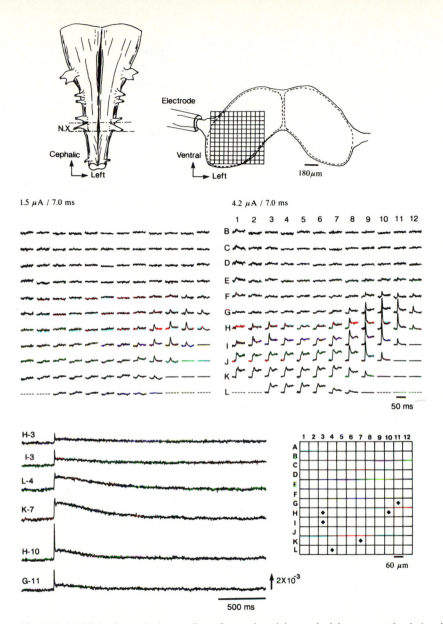

Fig. 21.4. Multiple-site optical recording of neural activity evoked by vagus stimulation in an embryonic chick brain stem slice preparation. The slice preparation was made by transverse sectioning of an 8-day-old embryonic brain stem at the level of the vagus nerve (as illustrated on the *left upper corner*). Positive square current pulses were applied to the right vagus nerve fibers by a suction electrode. The evoked optical signals were detected by the 12 × 12-element photodiode array positioned on the image of the *right side area* of the brain stem. The relative position of the photodiode matrix array is illustrated on the *upper right corner* of the optical recording. The recordings were made with 1.5 μA/7.0 ms (for the *left-hand* recording) and 4.2 μA/7.0 ms (for the *right-hand* recording) square current stimuli. Wavelength 702 ± 13 nm. Enlargements of the optical signals obtained with 4.2 μA/7.0 ms stimulus current from six different positions (H-3, I-3, L-4, K-7, H-10, and G-11: indicated by *solid diamonds* on the grid corresponding to the photodiode matrix array shown on the *right side*) are shown at the *bottom*. The recordings were made in a single sweep

5.1 Synaptic Fatigue

When a 0.1 Hz train of stimulation (8 μA/5.0 ms) was applied, the magnitude of the slow signal was gradually decreased, although the fast signal was not affected. The fast signal size was nearly constant, but the slow signal size decreased exponentially. It is likely that this effect reflects synaptic fatigue and, if so, it argues that the slow signal is intimately related to a postsynaptic potential. Fatigue of this type was produced more rapidly in younger embryos than in older ones [5].

5.2 Effects of External Ca^{2+}

Figure 21.5 shows the results of a series of experiments designed to examine the calcium dependence of the slow signal. The traces represent signals detected from four different positions in the dorsal surface of an 8-day-old embryonic preparation. In each row, the optical signal on the left was recorded in control Ringer's solution containing 1.8 mM Ca^{2+}, and the right traces show the effect of Ca^{2+}-free (no added Ca^{2+}) Ringer's solution. In the normal Ringer's solution, both fast and slow signals were evoked by an 8 μA/5 ms stimulating current. When 1.8 mM $CaCl_2$ was replaced by 1.8 mM $MgCl_2$, the slow signals were entirely eliminated, while the fast signals were decreased only slightly. The effects of changes in external calcium concentration were always reversible [5].

Furthermore, in a Ringer's solution containing Mn^{2+} or Cd^{2+}, the slow signals were entirely eliminated. While, the fast remained. These experimental results are also consistent with the assumption that the slow signal corresponds to a postsynaptic potential [5].

Control Ca^{2+}-free

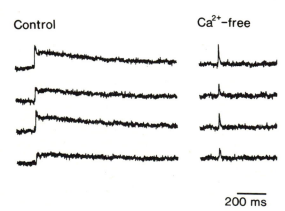

200 ms

Fig. 21.5. Optical signals recorded in the control (*left*) and a Ca^{2+}-free bathing solution (*right*). In the Ca^{2+}-free Ringer's solution, $CaCl_2$ was replaced by $MgCl_2$. The recording was made 20 min after the replacement of Ca^{2+} with Mg^{2+}. The signals from four different positions are represented. An 8-day-old preparation was used. Note that the slow signals were eliminated in the Ca^{2+}-free bathing solution

6 Evidence for Glutaminergic EPSPs

From the viewpoint of the ontogenesis of synaptic function, it is of interest to know which neurotransmitters participate in the generation of the slow component of the optical signal. Accordingly, we have examined the effects of several blockers of synaptic transmission on the slow signals.

Figure 21.6 shows the effect of kynurenic acid (1.2 mM), a blocker of glutamate-mediated excitatory synaptic transmission [17], on the evoked optical signals. In the control recordings, the slow signals are clearly visible. These slow signals were abolished in the presence of kynurenic acid (also see Fig. 7A). There was some recovery of the signals after kynurenic acid was rinsed off. Not surprisingly, the fast component of the optical signals was insensitive to kynurenic acid [5].

The glutamate receptor has been classified into four main subtypes, viz., NMDA (N-methyl-D-aspartate), quisqualate, kainate and 2-amino-4-phosphonobutyrate (APB) [17]. We also observed the effects of DL-2-amino-5-phosphono-valeric acid (2-APV: NMDA-receptor antagonist: [18]) and of 6-cyano-7-nitroquinoxaline-2, 3-dione (CNQX: non-NMDA-receptor antagonist [19]) on the postsynaptic slow-signal. As shown in Fig. 21.7, 2-APV depressed the later phase of the slow-signal, while CNQX reduced the initial phase of the slow-signal. The slow-signal was mostly abolished in the presence of both 2-APV and CNQX. These results suggest that the slow signals evoked by stimulation of the vagus nerve fibers represent a compound glutamate-mediated excitatory postsynaptic potential [5].

In order to examine the possibility that other chemical transmitters (such as adrenalin, acetylcholine, GABA and glycine) play significant roles in the synaptic physiology of the embryonic brain stem, we examined the effects of several other blockers. However, since we have not found significant effects of these blockers on the slow signals, it seems less likely that other transmitters are active in generating the slow signals [5].

7 Discussion

The results presented here demonstrate postsynaptic potentials from the early embryonic vertebrate brain stem at times close to the origin of synaptic function, using an optical technique for multiple-site recording of electrical activity.

In the present experiments, first, we must consider the following basic points:

(1) From histological evidence, it is likely that the vagal response area within the early embryonic brain stem includes many motor neurons and sensory nerve terminals and that glial cells might not be differentiated [20]. There is no detailed information concerning interneurons.

(2) The vagus nerve bundle contains both motor and sensory nerve fibers. Thus, in the present experiment, the applied stimulation was simultaneously

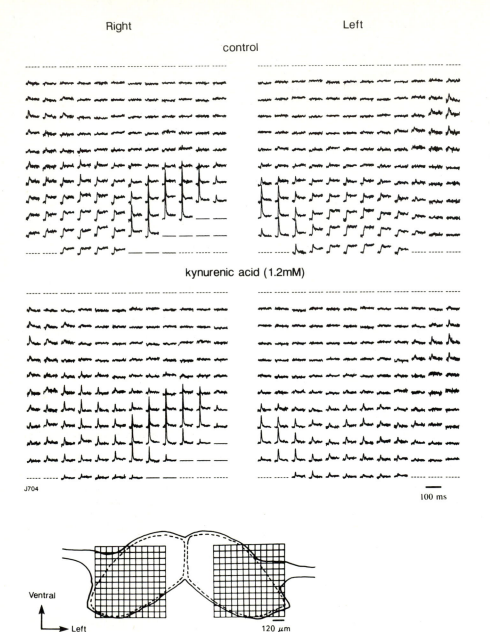

Fig. 21.6. Effects of kynurenic acid on the evoked optical signal. Signals were recorded simultaneously from 127-different positions in the *right-side* area (for the right vagus stimulation) and the *left-side* area (for the left vagus stimulation) of an 8-day-old preparation. The square current pulses of 6.0 μA/5 ms were applied to the right and left vagus nerves. The upper recordings are control, and the lower recordings were made in the presence of kynurenic acid. The recording was carried out 15 min after the addition of kynurenic acid (1.2 mM final concentration) to the bathing solution. The relative position of the photodiode array on the image of the preparation is shown at the *bottom*

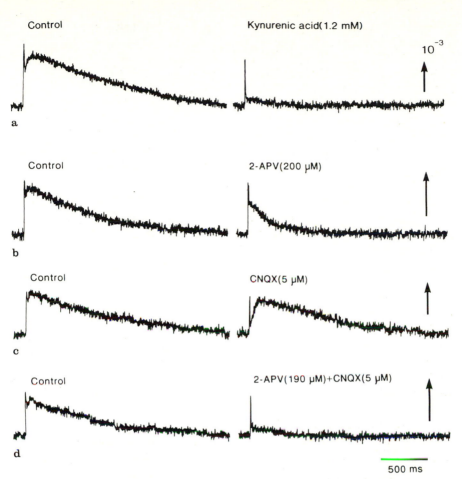

Fig. 21.7a–d. The effects of kynurenic acid, 2-APV and CNQX on the post-synaptic slow-signals in 8-day-old brain stem slices. **a** kynurenic acid (1.2 mM) was applied; **b** 2-APV (200 μM) was applied; **c** CNQX (5.0 μM) was applied and **d** 2-APV (190 μM) and CNQX (5.0 μM) were applied together. The recordings were made about 20 min after application of the drugs. Recordings A, B, C and D were obtained from four different preparations

orthodromic (for the sensory nerve fibers) and antidromic (for the motor nerve fibers).

(3) In our optical recording system, each element of the photodiode array detects optical signals from many neurons and processes. Thus, the signal size is proportional to the magnitude of the membrane potential changes in each cell and process and to the number and membrane area of these elements within the field viewed by one photodiode [21, 22].

In the brain stem preparations used in the present experiments, it is reasonable to assume that the fraction of the receptive field occupied by the nerve terminals of the sensory neurons is much smaller than that of the motor neurons. Therefore, we may suppose that the fast signal is mainly generated by the antidromic action potential in the motor neurons and that the component of the signal related to the orthodromic action potentials of the sensory nerve terminals is relatively small.

The results shown in Figs. 21.5, 21.6 and 21.7 suggest that the slow signal reflects the postsynaptic potential. The duration of the slow signal was longer than 500 ms. This observation suggests that the postsynaptic potential in the early embryo is very slow, and this may be characteristic of embryonic synaptic transmission.

Because the optical receptive field of a single photodiode was $56 \times 56 \ \mu m^2$ in the present experiments, we have not attempted to analyse the optical signals at the single cell level, and the spatial resolution is not presently fine enough to allow a substantial fraction of the postsynaptic neurons to be identified. However, a substantial area of postsynaptic activity generated in the vagus nerve-related sensory nucleus during early phases of development. It should be pointed out that this sensory area is identified functionally [6].

In the nucleus of solitary tract (*Nucleus tractus solitarii*) of adult vertebrates, *L*-glutamate has been shown to be a chemical transmitter related to the baroreceptor reflex [23, 24]. Thus, there is strong reason to suggest that the postsynaptic response area detected in the present experiment corresponds to the nucleus of solitary tract and that in this nucleus, functional synaptic connections are first formed at the 7- to early 8-day stage of development. No synaptic potential-related optical signals were detected from early 6-day-old embryonic brain stem [2, 3, 6].

Acknowledgement This work was supported by grants from the Ministry of Education, Science and Culture of Japan, the Suzuken Memorial Foundation, the Brain Science Foundation of Japan, Kowa Life Science Foundation and the Inoue Foundation of Science.

8 References

1. Sakai T, Hirota A, Komuro H, Fujii S, Kamino K (1985) Developmental Brain Research. 17: 39
2. Kamino K, Katoh Y, Komuro H, Sato K (1989) Journal of Physiology 409: 263
3. Kamino K, Komuro H, Sakai T, Sato K (1990) Neuroscience Research 8: 255
4. Sakai T, Komuro H, Katoh Y, Sasaki H, Momose-Sato Y, Kamino K (1991) Journal of Physiology 439: 361
5. Komuro H, Sakai T, Momose-Sato Y, Hirota A, Kamino K (1991) Journal of Physiology 442: 631
6. Momose-Sato Y, Sakai T, Komuro H, Hirota A, Kamino K (1991) Journal of Physiology 442: 649
7. Cohen LB, Salzberg BM (1978) Reviews of Physiology Biochemistry and Pharmacology 83: 35
8. Salzberg BM (1983) Optical recording of electrical activity in neurons using molecular probes. In: Barber JL (ed) Current Methods in Cellular Neurobiology. vol. 3 Electrophysiological Techniques, John Wiley, New York, 139
9. Cohen LB, Lesher S (1986) Optical monitoring of membrane potential: Methods of multisite optical measurement. In: De Weer P, Salzberg BM (eds) Optical methods in cell physiology, Wiley Interscience, New York, p 71

10. Grinvald A, Frostig RD, Lieke E, Hildesheim R (1988) Physiological Reviews 68: 1285
11. Kamino K (1991) Physiological Reviews 71: 53
12. Fujii S, Hirota A, Kamino K (1981) Journal of Physiology 319: 529
13. Kamino K, Hirota A, Fujii S (1981) Nature 290: 595
14. Ross WN, Salzberg BM, Cohen LB, Grinvald A, Davila HV, Waggoner AS, Wang CH (1977) Journal of Membrane Biology 33: 141
15. Salzberg BM, Grinvald A, Cohen LB, Davila HV, Ross WN (1977) Journal of Neurophysiology 40: 1281
16. Gupta RK, Salzberg BM, Grinvald A, Cohen LB, Kamino K, Lesher S, Boyle MB, Waggoner AS, Wang CH (1981) Journal of Membrane Biology 58: 123
17. Foster AC, Fagg GE (1984) Brain Research Reviews 7: 103
18. Nelson PG, Pun RYK, Westbrook GL (1986) Journal of Physiology 372: 169
19. Yamada KA, Dubinsky JM, Rothman SM (1989) Journal of Neuroscience 9: 3230
20. Fujita S (1964) Journal of Comparative Neurology 122: 311
21. Orbach HS, Cohen LB, Grinvald A (1985) Journal of Neuroscience 5: 1886
22. Kamino K, Hirota A, Komuro H (1989) Advances in Biophysics 25: 45
23. Perrone MH (1981) Brain Research 230: 283
24. Talman WT, Granata AR, Reis DJ (1984) Federation Proceedings 43: 39

Subject Index

D. A. W. Wendisch

Acronyms and Abbreviations in Molecular Spectroscopy

An Encyclopedic Dictionary

1990. V, 315 pp. 10 figs. Hardcover DM 98,–
ISBN 3-540-51348-5

Acronyms are extensively used to name new spectroscopic methods. The uninitiated reader of research papers is often confused with terms like ISIS, NERO, TANGO. This dictionary gives correct definitions of acronyms frequently used in molecular spectroscopy and imaging methods, descriptions of physical effects, practical applications and references to the scientific literature for further reading. The bulk of the more than 450 acronyms explained are from NMR and MRI, IR, RAMAN and ESR methods. The dictionary is arranged alphabetically and indexes, e.g. a subject index, allow easy access to the information. This book is an invaluable source for the spectroscopist.

Springer-Verlag
Berlin
Heidelberg
New York
London
Paris
Tokyo
Hong Kong
Barcelona
Budapest

D. L. Andrews (Ed.)

Perspectives in Modern Chemical Spectroscopy

With contributions by numerous experts

1990. XII, 325 pp. 187 figs. 15 tabs.
Softcover DM 58,– ISBN 3-540-52218-2

This book provides a broad and up-to-date coverage of techniques in chemical spectroscopy. The subject is treated very much with the practitioner in mind, and is principally aimed at those in industry.
The book is based on a week-long course with the same name which has been held annually at the University of East Anglia since 1979. It is a textbook providing its reader with a perspective on applied spectroscopy which, by reference to numerous examples, will illustrate the strengths and weaknesses of each of the major techniques now routinely available. It also provides experienced spectroscopists with an introduction to some of the most recent techniques. Emphasis is placed on practical aspects and on instrumentation, sample handling and the interpretation of spectra.

Springer-Verlag
Berlin
Heidelberg
New York
London
Paris
Tokyo
Hong Kong
Barcelona
Budapest